# Making Global Health Care Innovation Work

# Making Global Health Care Innovation Work
## Standardization and Localization

Edited by Nora Engel, Ine Van Hoyweghen, and
Anja Krumeich

palgrave
macmillan

MAKING GLOBAL HEALTH CARE INNOVATION WORK
copyright © Nora Engel, Ine Van Hoyweghen, and Anja Krumeich, 2014

First published in 2014 by PALGRAVE MACMILLAN® in the United States—a division of St. Martin's Press LLC, 175 Fifth Avenue, New York, NY 10010.

Where this book is distributed in the UK, Europe and the rest of the world, this is by Palgrave Macmillan, a division of Macmillan Publishers Limited, registered in England, company number 785998, of Houndmills, Basingstoke, Hampshire RG21 6XS.

Palgrave Macmillan is the global academic imprint of the above companies and has companies and representatives throughout the world.

Palgrave® and Macmillan® are registered trademarks in the United States, the United Kingdom, Europe and other countries.

ISBN: 978-1-137-45602-1

Library of Congress Cataloging-in-Publication Data

Making global health care innovation work : standardization and localization / edited by Nora Engel, Ine Van Hoyweghen, and Anja Krumeich.
        p. ; cm.
    Includes bibliographical references and index.
    ISBN 978-1-137-45602-1 (hardcover : alk. paper)
    I. Engel, Nora, editor. II. Hoyweghen, Ine Van, editor. III. Krumeich, Anja, editor.
    [DNLM: 1. Biomedical Research—ethics. 2. Biomedical Research—standards. 3. Developing Countries. 4. Socioeconomic Factors. 5. World Health. W 20.5]
    R850
    610.72'4—dc23
                                                          2014017360

A catalogue record of the book is available from the British Library.

Design by Amnet.

First Edition: October 2014

10 9 8 7 6 5 4 3 2 1

# List of Figures/Tables

# Contents

# Preface

The chapters that form this edition have been selected from the best master theses from the last three years of the Masters of Global Health at Maastricht University. Students from various disciplines, backgrounds, and corners of the globe meet to study how trends in globalization impact the health of communities across the globe. Throughout the year, they learn how to critically analyze the latest approaches, policies, and innovations in biomedicine, health care, and health promotion and to understand how these interact with local and global conditions that impact life, work, and health in a myriad of ways. Functioning in international and multidisciplinary settings is yet another skill students acquire as they work on real life assignments in collaboration with peers from partner universities in India (Manipal University), Thailand (Thammasat University), and Canada (McMaster University). They collaborate in online, transcontinental study teams or when they take electives at one of the partnering universities. They also apply some of these theories, approaches, and methodologies to their own research studies. Some combine their educational and geographic backgrounds to explore topics that they had pondered for a long time; others plunge into new disciplines, geographies, and problems.

The overall theme of the book is the result of a unique focus on innovation for global health and questions of design, transferability, and implementation of global health innovations in the Global Health Master Program. These questions are addressed throughout the program, but are especially addressed in an elective track on implementing innovations that explicitly asks, in several modules, how to transfer innovations and how to make them work in different contexts. We approach these questions by using insights from innovation and implementation studies, Science and Technology Studies, and practical examples of global health innovations. Several of the authors conducted their research through these questions, puzzles, and approaches, while others brought in refreshing contrasting perspectives onto the questions of innovation, standardization, and localization in global health. As such, compiling this edition has been a learning process for everyone involved. We hope that this edition, with its specific outlook, offers an important contribution for an emerging field of social studies of global health.

A special thank you goes to the participants of the research and the staff at the institutions that hosted our students during their fieldwork and made these studies possible. Also, we would like to express our sincere appreciation of all the staff involved in the Global Health Program.

In preparing this edition, we were grateful for support from Hans Maarse and the colleagues at the Department of Health, Ethics, and Society, and especially Bart Penders for tips and tricks at the outset of this project and Hellen Heutz and Angelique Heijnen for administrative support. We are very grateful to Clodagh McLoughlin for copy editing the first manuscript of the book. We are also thankful for the enthusiastic support, encouragement, and helpful tips from the team at Palgrave Macmillan and the anonymous reviewers.

Lastly, we thank the authors and their coauthors for all of their hard work and time invested in contributing to this edition. We learned a lot from you, but above all it was fun to do this.

Nora Engel, Ine Van Hoyweghen, and Anja Krumeich,
Maastricht 2014

# INTRODUCTION

# Making Health Care Innovations Work: Standardization and Localization in Global Health

*Nora Engel, Ine Van Hoyweghen, and Anja Krumeich*

## Introduction

Global health is becoming an increasingly political, professional, and academic field of its own. New actors and forms of collaboration across borders have emerged to solve some of the world's most daunting public health problems. Central among these efforts are a multitude of global health care innovations (drugs, diagnostics, and vaccines as well as policy initiatives and service delivery strategies) designed, developed, and implemented by a range of actors across very different institutional and cultural settings. Underlying these innovations is the common assumption that universal solutions can be found and brought to scale if implementation challenges of different contexts are overcome (e.g., function in weak health care systems where infrastructure, staff, and capacities might be absent). Standardized drug delivery programs are such a universal solution that allows participants to use the same training modules, delivery strategies, technologies, and logistics across very different contexts. These programs promise efficiency and the ability to reach large populations quickly while generating comparable global data through the same reporting and recording guidelines and standards. These are clearly beneficial aspects of universal solutions, and the belief in such magic silver bullets persists; however, global health technologies and solutions are not easily transferred to different local contexts.

When making global health care innovations work across different settings, contexts, and social worlds, there is a central tension between standardization and localization. For example, a recently developed new standardized tuberculosis test, GeneXpert, which was heralded by the WHO to revolutionize tuberculosis control in clinics and health posts of high burden countries, did not easily function in varied settings. The test often needed additional staff, infrastructure, and financial support in order to work (Clouse et al. 2012). Similarly, standardized measurement indicators used in a clinical microbicide trial assume that there is a fixed reality that can be measured in a universal and standardized way. Yet the survey questions, such as how often participants had sexual intercourse, meant different things to different participants (it could involve meeting your partner or several rounds spread across an entire day) and also changed meaning over the course of the trial period. In other words, this standardized measurement indicator did not make sense in every locality, changed in response to the research, and led to discrepancies in reporting the number of sexual acts (Montgomery and Pool 2011).

This tension between universal standards and localized health care innovations is the focus of this book. The core argument is that in order to make health care innovations work across different contexts and settings, as is common in the global health field, we need to study how these innovations travel to other hospitals, organizations, institutions, countries, or regions. We need to map out how actors are making health care innovations "work." This book thereby makes an important and unique contribution to recent studies on global health. Some recent works have focused on introducing the emerging field of global health and include past and future contributions of social science research to interested students and researchers (Farmer et al. 2013; Nichter 2008). Others focused on specific topics such as biomedical research in Africa (Geissler and Molyneux 2011), critical anthropological research of global biomedicine (Lock and Nguyen 2010; Petryna 2009) and global health practices (Biehl and Petryna 2013), multidirectional learning opportunities between low- and high-resource settings (Crisp 2010), and the relationship between globalization, governance, and global health (Buse, Hein, and Drager 2009; Labonté et al. 2009; Lee 2003). With its focus on the transferability of health care innovations, the present book takes a novel approach to developing, evaluating, and assessing innovations in the field of global health.

## Health in a Globalized World

Global health is closely related to processes of globalization. Globalization has brought numerous challenges to health systems and to the health and well-being of citizens across the world. These challenges include the increasing inequity and poverty between countries as well as within countries and cities, the spread of infectious diseases, drug resistance, unhealthy lifestyles, and the global promotion of junk food. Government spending on health is decreasing due to global neoliberal policies (e.g., structural adjustment programs) and volatile trade and investment flows, while health risks increase due to global environmental change and global criminal activities, brain drain in the health sector, and global health workforce migration. Asymmetrical imbalances in global markets have consequences for access to food and pharmaceuticals. On a political level, national policy spaces to govern health care challenges are being contracted (e.g., by trade agreements), and global intellectual property regimes impact the development and access to drugs for marginalized populations or diseases of poverty (Buse, Hein, and Drager 2009; Labonté et al. 2009).

Along with these challenges, opportunities have emerged through globalization for solving "global health challenges". These opportunities include improvements in health care and sanitation, along with the general economic development of countries; globally coordinated governance frameworks that allow cooperation across national or regional boundaries; cooperation and mass storage of data through global communication technologies; globalized media as watchdogs and awareness creators; the empowerment of the disadvantaged; and the diffusion of knowledge and innovations across borders (Buse, Hein, and Drager 2009; Labonté et al. 2009). The potential to learn from experiences of providing health care in resource-constrained settings in poorer parts of the world provides hope, especially with regard to rapidly aging societies and exploding health care costs in high-income countries. For instance, actors in these countries are developing a range of innovations to overcome the lack of human resources, such as task shifting or mobile health initiatives (Crisp 2010). However, the potential of globalization to deliver substantial health benefits remains largely untapped (Buse, Hein, and Drager 2009). Innovations are central in coping with global health challenges, yet the spread of solutions and good ideas across borders is often only an afterthought when discussing health in a globalized world. The fears of the uncontrolled spread of diseases and infectious strains across borders remain center stage. Changing this outlook will require more attention to processes and modes of innovation to improve and strengthen those processes.

The notion of "global health," as used in this book, distinguishes itself from earlier notions of "international health" and "tropical medicine" mainly through its focus on interactions between local and global dimensions or determinants of health care challenges. This understanding is based on a realization that today's health problems are complex, based on a variety of factors that transcend disciplinary, geographical, political, institutional, and sectorial boundaries. Solutions to these problems need to cross these boundaries and understand health problems in their global and local dimensions at the same time; thus, national governments alone are unable to solve them. This understanding also means that global health problems are ubiquitous and not confined to specific geographic contexts, such as the so-called global south. Such an understanding renders earlier dichotomies of developed vs. underdeveloped, global south vs. global north, and ill vs. healthy obsolete. However, these dichotomies are often an inherent and usually unarticulated assumption of global health research and interventions, despite the existence of hunger, pockets of high infant mortality rates, and infectious disease outbreaks in the US and EU. Therefore, a risk of presenting old wine in new bottles exists. Many of the old routines and institutions have been merely relabeled, while old relationships, binary worldviews, and power structures are being maintained. For example, the relationship between what was formerly known as donor and recipient is now relabeled in terms of partners, networks, and partnerships in an increasing number of public-private partnerships (Buse and Walt 2000).

## Science, Technology, and Innovation for Global Health

There is growing interest and awareness among global health policymakers in the role that science and technology play in improving the well-being and health in people's lives. For example, the Commission for Africa convened by the UK government in 2005 recommended funding more research and development for medicine and vaccines. As part of a broader program of capacity building, it also recommended the commitment of $3 billion over ten years to develop centers of excellence in science and technology across African institutes (Commission for Africa, 2005). In the same line, a new generation of donor, philanthropic, and public-private initiatives has emerged and has attracted increased funding. Some examples of these new initiatives are heavily-funded global initiatives such as the Global Fund to Fight AIDS, TB, and Malaria; the Human Genome Diversity Project; and the GAVI Alliance (formerly the Global Alliance for Vaccines and Immunization), all of which have science and technology at their core. In addition, Bill Gates has begun targeting money in his foundation for 14 "grand challenges" for research in global health (Grand Challenges in Global Health). New hopes are encouraged through successes such as the polio eradication in India, supported through massively increased immunization and surveillance efforts by the government (Park 2014). Many stakeholders make claims about the promises and expectations that new technologies may have with respect to major health problems in a globalized world: new drugs, vaccines, diagnostics and infrastructure applications. Some stakeholders claim that major technological breakthroughs could solve longstanding health problems and tackle emerging disease outbreaks across the globe.

Many advocates of linking science, technology, and global health assume that the resulting innovations will be diffused across the globe through standardization. Current investments are justified by the prospect of "big hit" technologies that have potential for global scope, applicability, and the capacity to deliver these on a large scale. Policy debates are dominated by the beguiling idea that there are top-down technological fixes that act as "magic bullets" for today's global health problems. Scientists have become well-versed in making their cases to big funders along these lines, often promising more than is realistically deliverable. While such initiatives claim to be attuned to health improvement, problems and solutions are generally framed in universalized terms—applicable anywhere, anytime. The nature of health problems is assumed to be broadly similar across vast areas, so it is assumed that standardized solutions can be transferred and applied at scale (Leach and Scoones 2008).

Despite the potential of such "big hit" technologies, there are many examples of failed innovation transfers in global health. Some innovations have not traveled well, have not been used, have broken down, have not been applicable, or have not been available in all contexts. These include disease eradication programs that are met with public resistance because the programs are perceived as inappropriate. Such was the case of the global polio eradication initiative in Nigeria, the tetanus toxoid campaigns in Uganda and Cameroon, and the American and French Ebola control measures in Gabon in 1995–96. The control measures in Gabon were perceived as so inappropriate and offensive by villagers that the following international responses to another outbreak were met with fierce, locally-armed resistance (Leach and Scoones 2008). Also, bed nets to control vector-borne diseases do not always easily reach those in need but are instead diverted to the black market, are out of stock at primary health clinics, end up not being used for a particular group of children, or are used as fishing nets or wedding veils (Easterly 2006; ch. 5). Some innovations can even distort control efforts in the long run. For instance, vertical disease programs are targeted innovations that can be very effective but often draw resources away from general health services, distorting those services over time (Garrett 2007). Another example is the development of new drugs that can lead to the emergence of drug resistance if those innovations are misused; a lack of attention to nonbiomedical disease determinants and preventive measures can also occur if drugs are seen as magic bullets, threatening the public health control efforts in the long term (Atre and Mistry 2005).

This is not to deny the key role that innovations, such as drugs, prevention programs, diagnostics, and vaccines, play in solving global health challenges. However, these innovations may miss their targets if technological and program choices and strategies for promoting innovation uptake are not adapted to local contexts (Leach and Scoones 2006). Thus, innovations are not always beneficial. The implications are that policymakers and funders also need to rethink how innovations are evaluated and scaled-up, as well as how investment priorities are decided.

## Transferability of Health Care Innovations

The idea that universal health care solutions can be designed and then transferred to different contexts is based on an approach of innovation diffusion (Greenhalgh et al. 2004; Rogers 2003). These authors try to define conditions and determinants for the successful process and content of innovation, including diffusion, dissemination, and implementation. If innovations fail to be transferred successfully, the context and actors are blamed for not implementing the innovation properly based on lack of capacity, infrastructure, training, rigor, or adherence to protocols (Ogden 1999). For example, literature in health sciences dealing with the implementation of guidelines has examined the potential gap between best practice standards and real-life practices, proposing measures to close this gap (Bhattacharyya, Reeves, and Zwarenstein 2009). Implicit in these frameworks is a linear approach that assumes innovations and technology can be developed separately and only be implemented later, at which point they remain unchanged.

Science and technology studies (STS) scholars have argued that the form and direction taken by science and technology are no longer seen as inevitable and monolithic, awaiting discovery in nature and then transferred to society; they are instead increasingly recognized as being open to shaping by individual creativity, collective ingenuity, cultural priorities, institutional interests, and stakeholder negotiation. Science shapes society but science is also produced and shaped by societal factors (Bijker 2009; Jasanoff 2004; Latour and Woolgar 1979). Consequently, those societal aspects need to be taken into account when producing science and technology. Global health science initiatives that put forward "big hit" technologies do often not work precisely because they assume a linear approach of transferring science to society (Biehl 2013; Engel 2012; Michael and Rosengarten 2013; Müller-Rockstroh 2011). Conceptualizing the future of global health innovations is not just about building the hardware of R&D infrastructure and capacity, but it also requires developing the fundamental software of social relations among the many actors that are now involved and the different interests that shape science and technology agendas.

This understanding also qualifies a presumed controversy or dilemma between technological fixes and broader socioeconomic interventions that address the root causes of health problems and different conceptualizations of innovation: donor/externally-driven vs. locally developed. Social scientists in particular have questioned technological fixes over broader social and economic interventions that address global health problems; these scientists have also critiqued the largely US-driven conceptualization of technologies rather than local technological derivations or nontechnological, people-centered options (Singer and Baer 2007). Yet, this critical approach to technological determinism risks excluding technological factors at the extent of social ones (e.g., social determinism). Both of these approaches lack the eye for the mutual alignment between technology and society that is needed to make health care innovations work (Anderson 2006).

## Making Global Health Innovations Work: Understanding the Tension between Standardization and Localization

In the same vein, STS scholars have shown that standardization is not a matter of the straightforward implementation of a standard; it requires continuous work and renegotiation (Bowker and Star 1999; Lampland and Star 2009; Timmermans and Epstein 2010). The development of standards, such as guidelines, protocols, and policies, is characterized by continuous interaction between universal standards and local practices (Timmermans and Berg 2003). This means that universal standards do not always work in every locality and might mean different things (Anderson 2006; Lampland and Star 2009; Lock and Nguyen 2010; Petryna 2009). In making standards such as treatment guidelines for tuberculosis work, actors assess the role of guidelines in a particular situation; on that basis, actors recognize core recommendations of guidelines or go beyond guidelines. In this way, actors negotiate how standards (e.g., multi-drug resistant treatment guidelines in India) should be situated (Engel and Zeiss 2013). Both standards and the practices wherein they are applied are being shaped by this interaction.

How, then, do we make health care innovations work in different contexts? We need to understand what happens if health care innovations travel to another hospital, organization, institution, country, or region. We also need to know how actors are making health care innovations work and how they practice and enact technologies, such as drugs, vaccines, and diagnostics; but also policies, strategies, service delivery mechanisms, protocols, and guidelines. The production of innovations that work in different contexts needs to be supported by appropriate governance arrangements; in doing so, technology cannot be seen as an unproblematic input but needs to be critically examined in all its complexity.

A plea for a "slower race" has been introduced that takes into consideration the co-production of science and society through ideas such as citizen engagement and localizing science and biotechnology (Leach and Scoones 2006, 2008). Others have suggested more careful involvement of relevant forms of (local) expertise early in the innovation development process (Bijker, Bal, and Hendriks 2009; Engel 2012). This would imply assessing and negotiating participation of varied forms of expertise, allowing different local adaptations of standards based on an open negotiation of different trade-offs involved, and recognizing standardization as a dynamic change process (Engel 2012). These approaches mean that there needs to be much more attention and investment in processes of reflection, participation, consultation, and delivery and that there is investment in new forms of interdisciplinary research and education of new professionals able to bridge between different disciplines, localities, and publics.

This book takes these questions seriously; the different chapters highlight the work that is needed to make health care innovations "work" in different contexts. The chapters present this information in different ways: some by drawing on STS literature on social worlds, expertise, traveling of technologies, and standards and the making of science; others by reviewing cutting-edge science and policymaking in which questions of trade-offs between localization and standardization are center stage. This book connects innovations with local practices and offers a combination of interdisciplinary academic perspectives; it provides empirical insights into current global health challenges and discussions in relation to the tension of localization and standardization. The book thereby contributes to literature on STS by applying STS theories and approaches to an emerging and diverse global health field. It contributes to literature in health sciences by taking the tension between universal solutions and localization seriously in the way health care innovations are made, debated, and transferred to very different contexts.

The chapters address the disputes around technological fixes in multiple ways. Chapter 5, which discusses malaria, shows how an entire age group of children might go unprotected from a technological fix (bed nets) due to financial constraints. Chapter 4, which examines tuberculosis in Kenya, highlights the important role that community health workers play in making an externally-imposed technological fix (the tuberculosis control program) work in a particular community. The chapters on the practices of trials (chapters 1–3) explicitly address questions of externally-driven technologies and trial practices vs. locally-rooted practices of communication, benefit, and ethical expectations. The chapters in the second part of the book address these questions differently: Nutri-epigenetics (chapter 6) and environ-vaccinology (chapter 7) can be seen as technological fixes that suggest being able to circumvent some of these controversies, such as adapting vaccine development to different environmental contexts consisting of biophysical and sociocultural factors.

The chapters that form this book are based on a selection of master theses coauthored by thesis supervisors of the Global Health Master of Science Program at Maastricht University, the Netherlands. The majority of this edition's chapters are based on empirical data collected across a variety of geographical contexts (Ghana, South Africa, Kenya, India, Netherlands, Austria, global health policy arenas). While the chapters in the first part of the book focus on India and different countries of Sub-Saharan Africa (countries in the so-called "global south"), at least two of these countries, South Africa and India, are also countries labeled as emerging economies with health systems that offer world-class medicine alongside resource-constrained, unregulated, and, at times, poor-quality care in poor neighborhoods or regions. These countries struggle with problems similar to those in the EU/Americas along with problems traditionally thought to be limited to the global south. The chapter on folic acid supplementation (chapter 6) is focused on the EU with particular attention to the Netherlands and Austria. Furthermore, the chapters in part two focus on innovations at the global level, where applications to very different geographical contexts are envisioned and debated, including regions that are often labeled global south and north.

## Overview of the Book

The book is divided into two parts that explore the tension between standardization and localization in health care innovations in different ways. In part one, the authors examine how actors deal with this tension and make a particular standard work in a local context, revealing frictions and challenges (e.g., informed consent in clinical trials, community-based tuberculosis delivery strategies, insecticide-treated bed nets). The chapters in the second part of the book revisit debates around contested trade-offs in global health innovations that come from setting standards for research and development (i.e., patents) and review new research fields that promise to circumvent the dilemma of adapting technologies to individual bodies and local contexts (e.g., environ-vaccinology and epigenetics).

### Putting Standards to Use in Different Contexts

The five chapters that make up the first part of this book examine issues involved in transferring or transporting knowledge, expertise and standards, ethical practices and principles to another setting or context. The first three chapters do so by investigating practices of conducting clinical trials and ethical standards of informed consent or payments/benefits.

Akrong, Horstman, and Arhinful describe an exploratory study on local perspectives of clinical trials (participation) conducted in Ghana. The underlying meanings and implications accompanying respondents' ideas are discussed in relation to informed consent—considered a core pillar and standard of ethical biomedical research practice. Questions have been raised about the validity of obtaining "truly" informed consent from "vulnerable" populations in lower-income countries. This study found that respondent perceptions and understandings of clinical trials were tied to the ideas of local community benefit and responsibility and were (re)shaped throughout the trial process.

The chapter by van Alphen, Engel, and Vaz focuses explicitly on challenges to informed consent in a local trial setting in India. Based on the perspectives of involved actors, the study identifies challenges for informed consent processes such as literacy, language, poor medical awareness, therapeutic misconception, education, socioeconomic status, and doctor-patient rapport. Based on these results, the chapter questions the extent to which universalization of global health ethics has resulted in successful informed consent and calls for indigenized bioethics.

The chapter by Zvonareva and Engel tackles another standardized aspect of global clinical trial conduct: the controversial issue of payments to clinical research participants. By analyzing views of actual and potential research participants in South Africa on benefits in clinical research, the study highlights contradicting ethical expectations of trial participants and bioethical models of benefits and payments. Informants viewed clinical research as an activity bringing better health to host communities and did not expect money for participation. They were aware of the risks of medical experimentation magnified by limited accessible medical care, yet they indicated payments had little effect on their decision-making with regard to participation.

The two remaining chapters in the first section of the book explore the standards that are transferred as part of disease control programs. The chapter by Nguyen Thi Mai and Engel examines the work that community health workers do to make a standard disease control program (in this case tuberculosis) function in a community in Kenya. The results provide insights into the work and position of community health workers in the tuberculosis program. They engage in multiple roles and continuously cross the boundaries between the world of patients and health professionals, acting as mediators between the standard and the individual. This hybrid position involves complicated relations between different actors in a disease control program yet seems crucial to its success.

Eelens and Meershoek's study on the consequences of malaria control campaigns focuses on children under five and pregnant women in Kenya. The authors show that in situations of scarce financial resources, children ages 5–14 are more likely to sleep under a damaged net, with their treatment often being delayed or based on herbal remedies or painkillers. Here, a particular standardized technology (insecticide-treated bed nets) does not work in the everyday context of a particular age group and community. More involvement of user perspectives when designing global health standards is needed.

### Redesigning Standards and Making Public Health Trade-offs

The second part of the book explores how to develop standards that fit different contexts and ways to deal with public health trade-offs while doing so. The four chapters in this section explore these questions for very different fields. They do so by reviewing state-of-the-art literature, exploring perspectives of experts and policy translation in newly emerging research fields (environ-vaccinology or epigenetics), and by revisiting debates around contested trade-offs in global health innovations that emerge when setting standards for research and development protection (i.e., patents) and examining some of the proposed solutions to these challenges (e.g. prizes).

In these discussions, some authors take technology itself for granted and argue that the receiving context needs to be adapted; contrary to STS who do not take technology for granted and argue that technology needs to be changed as well. The first two chapters in part two seem to suggest a third solution by discussing two technologies or fields of new scientific practice: epigenetics and environ-vaccinology. These two fields promise to circumvent the tension between universal standards and local adaptation by being a standard/technology that will be adapted locally. Nutri-epigenetic knowledge allows the adaptation of universal standards (such a folic acid supplementation) to personal susceptibility of specific population groups; environ-vaccinology allows the adaptation of vaccine development to different environmental contexts consisting of biophysical and sociocultural factors.

Verhagen, Brand, and Ambrosino discuss how nutri-epigenetic knowledge concerning folic acid supplementation gets translated into present public health policy; and which stakeholders, factors, and relations influence this translation. In their chapter on local social confounders of vaccine response, Ling, Brand, and Ambrosino examine current state-of-the-art literature and expert opinions on the relationship between the social environment and responses to vaccines. They explore how such knowledge could be translated into public health policy by adapting vaccine development and strategies to local needs. Both of these chapters do not problematize technology but suggest a particular technology (epigenetics or environ-vaccinology) to circumvent the dilemma between local adaptation and standardization in health care solutions.

The last two chapters discuss the public health impact of setting contested standards and how to avoid detrimental impacts. Intellectual property protection and patents are particularly contested standards for research and development in the pursuit of innovations. Patents, as we consider them here, are a result of standard setting in research and development which might be caused by an entrenched practice of protecting utilization. This quest for standards is usually seen as a way to harmonize practices. These standards often result in global health trade-offs with, at times, detrimental impact on health. Uwland and Townend examine relevant literature and interviews with stakeholders to determine how international intellectual property protection influences research and development of new drugs for so-called neglected tropical diseases (NTDs) and why, despite patent protection, very few drugs have been developed in recent years. In a similar vein, Murray and Townend go on to discuss the feasibility of "prizes" as an alternative to the current incentive of patent protection for innovation and who should be involved when designing alternative global health standards.

With its focus on transferability of health care innovations in global health and by linking empirical, theoretical, and normative concerns related to global health innovations, the book provides a unique contribution to the emerging field of global health and to developing, evaluating, and assessing health care innovations. This information is relevant to policymakers and practitioners as well as to scholars in STS and innovation studies. Learning does increasingly happen in a multitude of ways, and old ideas of technology and knowledge transfer from the global north to the global south are outdated, inadequate, and have proven to fail. Yet these ideas are likely to continue to be prevalent in the thinking and operation of some actors and mechanisms in the global health field. We need to continue to identify, problematize, and criticize those ideas.

## References

Anderson, W. 2006. *The Collectors of Lost Souls. Turning Kuru Scientists into Whitemen.* Baltimore: Johns Hopkins University Press.

Atre, S. R., and N. F. Mistry. 2005. "Multidrug-Resistant Tuberculosis (MDR-TB) in India: An Attempt to Link Biosocial Determinants." *Journal of Public Health Policy* 26(1): 96–114.

Bhattacharyya, O., S. Reeves, and M. Zwarenstein. 2009. "What Is Implementation Research?: Rationale, Concepts, and Practices." *Research on Social Work Practice* 19(5): 491–502.

Biehl, J. 2013. "When People Come First: Beyond Technical and Theoretical Quick-Fixes in Global Health." In *When People Come First: Critical Studies in Global Health*, edited by J. Biehl and A. Petryna, 101–130. Princeton, Woodstock: Princeton University Press.

Biehl, J., and A. Petryna, A., eds. 2013. *When People Come First: Critical Studies in Global Health*. Princeton, Woodstock: Princeton University Press.

Bijker, W. E. 2009. "How Is Technology Made?—That Is the Question!" *Cambridge Journal of Economics* 34(1): 63–76.

Bijker, W. E., R. Bal, and R. Hendriks. 2009. *The Paradox of Scientific Authority. The Role of Scientific Advice in Democracies*. London and Cambridge, MA: MIT Press.

Bowker, G., and S. L. Star. 1999. *Sorting Things Out. Classification and Its Consequences*. Cambridge, MA: MIT Press.

Buse, K., W. Hein, and N. Drager. 2009. *Making Sense of Global Health Governance - A Policy Perspective*. Basingstoke: Palgrave Macmillan.

Buse, K., and G. Walt. 2000. "Global Public-Private Partnerships: Part I—a New Development in Health?" *Bulletin of the World Health Organization*, 78(4): 549–561. Accessed June 15, 2007 http://www.scielosp.org/scielo.php?script=sci_arttext&pid=S0042-96862000000400019&nrm=iso.

Clouse, K., L. Page-Shipp, H. Dansey, B. Moatlhodi, L. Scott, J. Bassett, W. Stevens, I. Sanne, A. van Rie 2012. "Implementation of Xpert MTB/RIF for Routine Point-of-Care Diagnosis of Tuberculosis at the Primary Care Level." *S Afr Med J* 102(10): 805–807.

Commission for Africa 2005. *Our common interest. Report of the Commission for Africa*. Accessed June 18, 2014 http://www.commissionforafrica.info/wp-content/uploads/2005-report/11-03-05_cr_report.pdf.

Crisp, N. 2010. *Turning the World Upside Down. The Search for Global Health in the 21st Century*. London: Royal Society of Medicine Press.

Easterly, W. 2006. *The White Man's Burden: Why the West's Efforts to Aid the Rest Have Done So Much Ill and So Little Good*. Oxford New York: Oxford University Press.

Engel, N. 2012. "New Diagnostics for Multi-Drug Resistant Tuberculosis in India: Innovating Control and Controlling Innovation." *BioSocieties* 7(1): 50–71.

Engel, N., and R. Zeiss. 2013. "Situating Standards in Practices: Multi Drug-Resistant Tuberculosis Treatment in India." *Science as Culture*: 1–25.

Farmer, P., J. Y. Kim, A. Kleinman, and B. Matthew, eds. 2013. *Reimagining Global Health: An Introduction*. University of California Press.

Garrett, L. 2007. "The Challenge of Global Health." *Foreign Affairs* 86(1). Retrieved from http://www.foreignaffairs.com/articles/62268/laurie-garrett/the-challenge-of-global-health.

Geissler, P. W., and C. Molyneux, eds. 2011. *Evidence, Ethos and Experiment. The Anthropology and History of Medical Research In Africa*. New York, Oxford: Berghan Books.

Grand Challenges in Global Health. www.grandchallenges.org

Greenhalgh, T., G. Robert, F. Macfarlane, P. Bate, and O. Kyriakidou. 2004. "Diffusion of Innovations in Service Organizations: Systematic Review and Recommendations." *Milbank Quarterly* 82(4), 581–629.

Jasanoff, S., ed. 2004. *States of Knowledge. The Co-Production of Science and Social Order.* London, New York: Routledge.

Labonté, R., T. Schrecker, C. Packer, and V. Runnels. 2009. *Globalization and Health: Pathways, Evidence and Policy.* New York: Routledge.

Lampland, M., and S. L. Star, eds. 2009. *Standards and Their Stories. How Quantifying, Classifying, and Formalizing Practices Shape Everyday Life.* Ithaca: Cornell University Press.

Latour, B., and S. Woolgar. 1979. *Laboratory Life: The Construction of Scientific Facts.* Los Angeles: Sage Publications, Inc.

Leach, M., and I. Scoones. 2006. "The Slow Race: Making Technology Work for the Poor." Accessed April 17, 2014. http://www.demos.co.uk/files/The%20Slow%20 Race.pdf.

Leach, M., and I. Scoones. 2008. "Health Dynamics, Innovation, and the Slow Race to Make Technology Work for the Poor." *Global Forum Update on Research for Health* 5: 124–29.

Lee, K. 2003. *Globalization and Health, An Introduction.* London: Palgrave Macmillan.

Lock, M., and V. K. Nguyen. 2010. *An Anthropology of Biomedicine.* West-Sussex, UK: Wiley-Blackwell.

Michael, M., and M. Rosengarten. 2013. *Innovation and Biomedicine: Ethics, Evidence, and Expectations in HIV.* New York: Palgrave Macmillan.

Montgomery, C., and R. Pool. 2011. "Critically Engaging: Integrating the Social and the Biomedical in International Microbicides Research." *Journal of the International AIDS Society* 14(2): 1–7.

Müller-Rockstroh, B. 2011. "Appropriate and Appropriated Technology: Lessons Learned from Ultrasound in Tanzania." *Medical Anthropology* 31(3): 196–212.

Nichter, M. 2008. *Global Health: Why Cultural Perceptions, Social Representations, and Biopolitics Matter.* Arizona: University of Arizona Press.

Ogden, J. 1999. "Compliance Versus Adherence: Just a Matter of Language? The Politics and Poetics of Public Health." In *Tuberculosis: An Interdisciplinary Perspective,* edited J. M. Grange and J. Porter, 213–234. London: Imperial College Press.

Park, M. 2014. "Global Health Success: India Certified Free of Polio." *CNN.* Accessed March 27. http://edition.cnn.com/2014/03/27/health/india-polio-3-years/.

Petryna, A. 2009. *When Experiments Travel: Clinical Trials and the Global Search for Human Subjects.* Princeton, NJ: Princeton University Press.

Rogers, E. 2003. *Diffusion of Innovations.* London: Palgrave Macmillan.

Singer, M., and H. A. Baer. 2007. *Introducing Medical Anthropology: A Discipine in Action.* Lanham, Plymouth: AltaMira Press.

Timmermans, S., and M. Berg. 2003. *The Gold Standard. The Challenge of Evidence-Based Medicine and Standardization in Health Care.* Philadelphia: Temple University Press.

Timmermans, S., and S. Epstein. 2010. "A World of Standards but Not a Standard World: Toward a Sociology of Standards and Standardization." *Annual Review of Sociology,* 36: 69–89.

# PART I

*Putting Standards to Use in Different Contexts*

# CHAPTER 1

# Informed Consent and Clinical Trial Participation: Perspectives from a Ghanaian Community

*Lloyd Akrong, Klasien Horstman, and
Daniel K. Arhinful*

## Introduction

Clinical research continues to expand globally. Lower-income countries have become popular destinations for research institutions and pharmaceutical companies in which to conduct clinical trials (Ballantyne 2010; Emanuel et al. 2004; Glickman et al. 2009). To illustrate, Ghana, a country relatively new to clinical research (Ogutu et al. 2010; Ghana-Michigan Collaborative 2010), reported having thirteen registered clinical trials in operation in 2012, according to the government's Food and Drug Board that regulates this work (Food and Drug 2012). Of these trials, twelve were sponsored by non-African institutions, with the remaining one funded by a Tanzanian research institution (Food and Drug 2012).[1] The pharmaceutical industry is aware that running trials in the lower-income countries is more cost effective than in higher-income countries (Petryna 2007; Schuklenk 2010). In lower-income countries like Ghana, participants are often targeted as preferable, "treatment-naïve" trial candidates, meaning individuals who have not been exposed to drugs that could potentially interfere with clinical trial results (Frimpong-Mansoh 2008; Mbuagbaw et al. 2011; Petryna 2007). Although the financial compensation given to lower-income country participants in clinical trials may be higher than what they typically can expect to receive in annual income, participants from low-income countries still receive relatively low compensation compared to their counterparts in wealthier nations

(Glickman et al. 2009; Brody 2002). This creates an economically beneficial arrangement for the pharmaceutical industry; lower research costs allow for increased profits margins (Petryna 2009). As various reports have shown, international clinical research is thriving financially and experiencing increasing revenues (Gatter 2006; Moses et al. 2005). According to PharmaTimes, revenues from the clinical trials market are expected to surpass $65 billion by 2021 (Mansell 2011). As the demand for pharmaceutical drugs and interventions grows, more international clinical research will be needed to meet this demand; consequently, greater numbers of global citizens will be needed to fill positions as trial participants in the various phases of research.[2]

The growing need for trial participants, combined with several notable moral atrocities that have plagued clinical research in Europe and North America (i.e. the German Nuremburg experiments and the US Tuskegee syphilis study), have led many critics to call for stronger participant protection mechanisms in international clinical research (Lorenzo et al. 2010; Moreno 2007). They believe this will help lower risk and mitigate the potential for the exploitation of "vulnerable" populations through participation in biomedical research (Dixon-Woods et al. 2006; Resnik 2009; Shamoo and Resnik 2006). While acknowledging this need for improved governance of international clinical trials and ensuring safety for participants involved in them, recent literature has been critical as well toward traditional protectionist approaches such as those seen in the informed consent process, arguing that they tend to be too heavily focused on the individual (Miller and Boulton 2007) and one's right to autonomy. One of the main arguments against such approaches is that they do not adequately, if at all, take into account the wider social context in which research and decision making take place (Felt et al. 2009). This includes clinical research in impoverished areas that have weak health system infrastructures and a lack of resources. Increasingly, there have been calls for more robust ethical research frameworks against the background of the increased shift of clinical trials to low-income countries (Buchanan et al. 2008; Levitt and Zwart 2009). In constructing these ethical frameworks, there is a need to ensure that they are contextually sensitive and inclusive to the views and concerns of research participants and the general public. Thus there is a clear need to critically consider how protection mechanisms currently used in international clinical research are implemented, negotiated, and enacted, including the use of traditional informed consent.

Informed consent of individuals from low-income countries engaging in internationally sponsored clinical research has been problematized by a number of publications (Krosin et al. 2006; London et al. 2012; Mystakidou and Panagiotou 2009; Tindana, Kass, and Akweongo 2006; Van Loon and Lindegger 2009). Commentators have voiced concerns regarding the ability of citizens from low-income countries to comprehend trial purposes, the motives behind their enrollment, and the implications of recruitment practices used in international clinical research (Angell 1997; Ellis et al. 2010; Krogstad et al. 2010). However, despite the seemingly ubiquitous adherence to the informed consent process, there remains no consensus on how to address these issues.

## A Contextual Study of Informed Consent

Informed consent has been widely endorsed as the cornerstone of ethical research practice (Beauchamp and Childress 2009; Belmont Report 1979; Faden, Beauchamp, and King 1986) and, consequently, promoted and adopted in international clinical research. The communication and comprehension of specific trial information is considered essential for obtaining genuine informed consent—and achieving the ethical goals ascribed therein—as outlined in international research guidelines, including that of the Belmont report (Belmont Report 1979). However, questions have arisen as to the effectiveness and utility of the universally recognized standard informed consent process (CIOMS 2002) and its ability to be implemented cross-culturally, particularly in low-income settings (Marshall et al. 2006), with particular cultures, norms, values, and traditions. Proponents of this framework have argued in favor of its validity. Literature suggests that this simplified method of communicating prespecified information through informed consent ensures the adequate conveyance of essential trial information (Reynolds and Nelson 2007). During the informed consent encounter, investigators are expected to provide prospective participants with important information regarding research purpose, trial design, possible risks, and benefits associated with participating in the research, as well as participants' rights (CIOMS 2002). This approach would provide prospective participants with the appropriate interpretation and relevance of trial-related concepts in order to make rational, well-informed choices and provide the basis on which to create appropriate relationship dynamics between investigators and participants, as well as respecting the participants' right to autonomy (Reynolds and Nelson 2007). Critics, however, argue that traditional informed consent establishes unequal power relations between the investigator and participant whereby investigators' "prevailing norms, values, and system of expertise shape the field of choice" (Corrigan 2003, 789) available to the research participants. Miller and Boulton (2007) emphasize that the definitions and meanings of various research concepts, including those comprising informed consent, are fluid and change as society changes. In addition, critics like Felt et al. (2009) have rejected the "information paradigm" mechanism established within informed consent, claiming that it does not take into account the social and contextual realities influencing decision making while naively assuming that the communication of "specific information" can lead to rational decision making (Ehrich et al. 2007; Stoljar 2011). Works of authors including Benatar (2002) and Kuczewski and Marshall (2002) have also been critical of informed consent practices that have been decontextualized, highly regulated, and formalized in aspiring to achieve objective universality.

A major concern often expressed regarding the practice of informed consent is that a lack of education, illiteracy, and/or language issues, particularly with trial participants from low-income countries, lead to a compromised ability to understand the often complex and technical biomedical "jargon" presented in informed consent documents (Terranova et al. 2012; Tindana, Kass, and Akweongo 2006). This issue has been emphasized as a key issue in exploitative relationships (Shamoo and Resnik 2006). In such cases, individuals are said to be incapable of fully appreciating the research in which they are enrolled or their rights as a participant (Oduro et al. 2008). Another concern is that individuals from low-income countries with limited job-market opportunities and unreliable, constrained, or inadequate health care systems are more susceptible to the influence of a therapeutic misconception (TM)—described as when participants confuse the activities of clinical research with those of therapeutic care (Appelbaum, Roth, and Lidz 1982; Appelbaum et al. 1987; Chingono et al. 2008; Glickman et al. 2009). Influenced by the perceived social and contextual challenges of informed consent, ethical deliberations have reflected on participants' voluntariness, decision making, and understanding of research purposes (Klitzman 2012; Petryna 2007, 2009). When hindered, these capacities are said to compromise the quality and validity of informed consent procedures (Silverman 2011).

Given these critiques of traditional informed consent and the inherent information paradigm, authors have suggested the need to reexamine the notion of informed consent altogether (Felt et al. 2009; Sariola and Simpson 2011). They suggest individuals have various ways of "knowing and decision making" (Felt et al. 2009, 88) that consist of mechanisms that occur in much broader contexts than expected (Morris and Bàlmer 2006). For example, a recent South African study showed that decision making and perceptions of clinical trial benefits were grounded in local, moral traditions and cultural beliefs beyond what was communicated to them during informed consent (Zvonareva et al. 2013). Additionally, Lock and Nguyen (2010) have described how pregnant women's decision making as part of biomedical research was made complex by participants' reluctance to share information with investigators; they were concerned that their local beliefs about the role the "spirit world" played in health could have negative consequences for their continued involvement.

## Designing the Methodology for Exploring Informed Consent in Ghana

This study explores the experiences and perspectives of antimalarial vaccine trial participants (n = 15) as well as individuals that consented to participate in the trial but were not able to follow through with participation for various reasons (n = 5).[3] The study examined their perceptions of trial information, participation, and research purposes. This chapter provides insight on the impact of informed consent as a protectionist mechanism in international clinical research by describing participants' experiences and viewpoints on issues associated with participation and exploitation. Combining ethnography and grounded moral analysis (Dunn et al. 2012), the first author (LA) conducted a qualitative research study that examined the knowledge and understanding of healthy participants involved in a phase I, placebo controlled study (from here on referred to as "the trial"); this study tested the safety and immunogenicity of an experimental antimalarial vaccine. The study was conducted at a research institution in the capital city of Accra, Ghana, West Africa. The trial was sponsored by a US institution and was described as the "first ever phase I vaccine trial" in the region, according to the research institute's staff (Personal Communication May 2012). In this study, there was a desire to collect experiences and perspectives of clinical trial conduct and participation from a wide array of groups and individuals. This group included individuals who were in favor of participation and those that were possibly more critical or skeptical of participation. In addition to the actual trial participants included in this study, it was acknowledged that individuals who refused personal participation invites could have provided more critical perspectives on participation and clinical trials conducted in the Ghanaian setting. However, the trial's recruitment strategy was not designed in such a way that individuals would be personally invited to participate. Therefore, we included individuals who consented to participate but then either withdrew or were not selected to participate. Trial recruitment was designed so that individuals who themselves expressed interest in trial participation would contact the research institute to arrange an appointment to attend an initial information and consent session. At this session, they would be provided with relevant trial information to review, take home, and then return at a later date to clarify any questions arising from that information.

Data was collected using document review and individual interviews with open-ended questions. Institute staff helped arrange contact with respondents. After contact was made, respondents were asked if they were interested in participating. Twenty out of twenty-one responded favorably. This study aimed to get a diverse sample of those who applied to participate in the study and either failed the screening, completed participation, or withdrew from participation. Interviews were conducted (in English) with 20 university students between 20- and 31-years-old. According to staff at the research institute in Accra, the sample group was characteristic of individuals found in many of the diverse communities throughout Ghana. Out of these 20 participants, 15 were participants in the malaria vaccine trial, 3 failed the required pretrial medical screening, one withdrew before the actual study began, and one was advised to withdraw by trial doctors after he experienced a fainting episode during a scheduled trial visit. Data collection and analysis were continuously performed throughout the study. This provided respondents the opportunity to comment and provide feedback on the accuracy of data interpretation. The other authors (KH and DA) contributed to data analysis and interpretation, enriching the process by drawing on their individual expertise: DA's extensive knowledge of the local research setting and Ghanaian culture; KH's training in philosophy and public health. Respondents were asked about research purposes, influences for trial participation, expectations, and their understanding of trial-related information. It was intended that more insight be provided on how informed consent was perceived, the importance associated with trial concepts, and how knowledge was translated and received. Initially, a semi-structured interview guide was developed based on themes identified through a literature review. During the interviews, respondents were encouraged to elaborate on these themes and/ or offer their own ideas on significant themes. As new themes were identified, the interview guide was reassessed and modified to reflect emphasized topics. In the next section, respondents' awareness and perspectives on clinical trials and their conduct, as well as the phases of clinical research, are discussed. Following this, personal considerations for research participation, views of the trial's purpose, and comprehension of research terms and concepts are explored. In the conclusion, the findings of this study are discussed against the background of wider debates presented in literature.

## Awareness of Clinical Trials and Research Phases

Several published studies, including Ekouevi et al. (2004) and Tindana, Kass, and Akweongo (2011), claim that lack of insight about clinical research activities is due to illiteracy or the absence of education. The majority of respondents in this study, however, were well-educated masters and undergraduate students. Still, in almost all cases, respondents reported having very little to no insight into the clinical trial research process. During the interviews, all respondents reported that, prior to the trial and attending the information sessions, they had no previous experience with clinical trials and had never heard of any trials being previously conducted in the area. One respondent mentioned that it was not until his brother and a friend told him about the trial that he developed an interest in participation:

*My brother works here. He works at another building (at the research institution). He told me that there was a clinical trial. And a friend also recommended it.* (28-year-old male, economics student)

When asked if the idea of trial participation ever occurred to him before being informed of the malaria trial, he gave a reply that applied to most respondents:

*No, actually not ... No, I didn't consider it. This was my first [experience].* (28-year-old male, economics student)

In general, there was little awareness or interest shown by respondents in regards to the current phase of the research trial. Therefore, it was not surprising that only one respondent independently brought up the concept of trial phases without being prompted. While the respondent demonstrated awareness of research phases, she incorrectly assumed the phase of research in which the vaccine trial was performed, which in fact was stated on the information sheet as a Phase I:

*I think it was the second phase ... everything was on the consent form they gave.* (27-year-old female, school of public health student)

Interestingly, when another respondent was asked if she knew in which research phase they were participating, she responded matter-of-factly:

*No, and I didn't ask.* (21-year-old female, English and political science student)

The tone of the given response suggested this was not viewed as critical information necessary for participation, and the respondent seemed surprised to be asked this question. This observation raises the question as to whether the lack of importance placed on research phases was related to not being aware of the concept of trial phrases or to having incorrect assumptions about its function. To probe further, a selection of respondents were asked to describe what they thought was meant by research trial phase I, phase II, phase III, and phase IV. Most were hesitant to venture a guess, stating they did not know or were not sure. The one respondent that was willing to speculate suggested:

> *From a hypothetical guess, I guess maybe a phase I, it's been done somewhere with a group of volunteers, then there is another group and then there is us, the third group.* (31-year-old male, psychology student)

The uncertainty of concepts and processes involved in clinical research was observed as a trend among a number of respondents in the study, regardless of whether they actually participated in the trial or not. A respondent who signed the consent form but withdrew from trial participation before it began was asked to elaborate on his knowledge of clinical trials and how the trial potentially played a role in drug development. He mentioned that, during his studies, he came across some literature on the clinical trial process, but he admitted that his ability to explain this process remained obscure. He then reiterated that he wished to participate in helping develop a vaccine by participating in the trial, emphasizing the importance of the work.

## Research Participation and Expectations: I Have My Reasons—it's Malaria!

Respondents' overall feelings toward clinical trial participation were encouraging and positive. The social value of what clinical trials as a public good could bring to those in local communities, as well as to the country as a whole, was heavily presented throughout this study. Due to the threat of malaria in Ghana, participating in research about the development of a malaria vaccine could help improve the country's health conditions significantly. This perception was established and supported by respondents who completed trial participation, those who failed to become actual trial participants, as well as the individuals who consented but withdrew from participation. A respondent who participated in the trial offered his reason for becoming involved with the research:

> *I was also interested in the aim of the research ... [looking] into the malaria vaccine.* (26-year-old male, political science and French student)

Further elaborating on the benefits of the trial for the local community, he shared that he felt malaria was one of the major health concerns for Ghana and other African countries. He expressed that by enrolling in this study, he felt he was actively doing something that could have a noticeable effect on the health of the community. This opinion was just as strongly shared by respondents who were not successful in their attempt at trial participation. A female respondent who hoped to be a participant but failed the pretrial medical screening spoke about her interest in participation. She also gave the impression that her interest was influenced by those around her, who also expressed enthusiasm about the opportunity:

> *I seen the posters around and then my friends also told me about it. Like they knew about it and I also saw it. And I was like wow, I'm interested.* (25-year-old female, history and theatre arts student)

Most respondents who completed participation in the trial were keen on becoming involved in future clinical trials. Although they had yet to receive news about the trial's success, they were hopeful that it was positive. Participation was seen as a social good and worthwhile activity by those who did participate; it was further explored whether respondents who consented but did not succeed in participating in the trial retained similar views on clinical research and future participation. Their inability to participate or complete participation in the trial did not seem to affect attitudes toward research or future participation. A respondent who failed the premedical trial screening enthusiastically responded about the prospect of future participation in research:

> *That I would like to participate? Oh yeah, like this um, how do I put it. Sometimes people go to the villages to give vaccinations, like Hepatitis B and all those stuff … I'm interested.* (25-year-old female, history and theatre arts student)

Her interest in joining research trials was a result of her association with activities resembling community vaccination programs that were run by government institutions and NGOs. It appeared her belief was that, much like these vaccination program activities, clinical research activities would benefit the health of local communities. Literature on therapeutic misconception (TM) suggests that if she was viewing research activities in a similar manner to that of a vaccination program, it may constitute a TM (Appelbaum et al. 1987; Goldberg 2011; Horng and Grady 2003), as long as one perceives the research activities as providing therapeutic benefit. Further probing found village vaccination programs and the trial were both viewed as "opportunities." The opportunities she highlighted included the chance to improve health through knowledge gained from the medical staff performing the trial or village vaccination programs.

Interestingly, another respondent revealed an important and possibly critical observation about how the research was initially presented to community members on the posters used in the advertising campaign:

*Well initially it wasn't like a clinical [trial]... it was like ok, volunteers needed for a clinical study.* (26-year-old male, environmental science student)

The fact that posters presented the research activities as a "study" and not a "trial" may have played a key role in how some individuals approached and negotiated the idea of research participation. Similar implications were suggested by the results of an American study examining patients' perceptions of medical research (Sugarman et al. 2012). The respondent who made the distinction between the work being introduced as a "study" rather than a "trial" was the only respondent to do so. Questioned further, he showed that the advertisement of a "study" did not provide a clear picture of what investigators intended to do. When asked about the significance of this distinction, he would not elaborate, although he did mention that it made him feel as though he needed more information from investigators.

Overall, respondents' comments point to the idea that general interest in participation was often directly related to how much the research focus could address local health concerns. The respondent who failed the pretrial medical screening argued that the research potentially having a positive impact in the community was important:

*What made me more interested was like the malaria. Cause like in Africa, in Ghana it's one of the sickness that kills a lot of people ... If I have to go to the hospital it's for malaria. You can easily contract malaria. I went, wow. If you are going to get this vaccine, it is going to help me; it is going to help friends, my family and then most people as Ghanaians.* (25-year-old female, history and theatre arts student)

Personal experiences with malaria corresponded with a strong desire to be a part of the trial. Investigators conveyed to participants that they may not receive the actual "experimental" vaccine and some would get a placebo (a neutral salt solution). Interviews demonstrated that some respondents did not fully comprehend the relevance of a placebo. A few of those who did understand believed there was still a chance they would receive the vaccine and the protection afforded by it. This implied there was a belief that the experimental drug was established to be effective, raising questions of TM.

Attitudes toward participation in research that did not have the possibility of improving local health conditions as the primary goal were diverse, but generally more skeptical. Some participants were apprehensive about enrolling in research that would not benefit participants and their communities. On the other hand, some individuals of this opinion admitted that, with adequate precautions and increased safety measures (i.e. health insurance, ancillary and post-trial care), they would be willing to contribute. At the least, it was suggested they would be more likely to consider it. A trial participant stipulated the conditions of his participation in research with a nonlocal focus:

*The disease could still come here so I would still most likely do it, but would need more health care compensation.* (30-year-old male, senior secondary graduate)

His concern was that he be provided health care compensation rather than financial compensation. It should be noted that participants in the trial were not financially compensated for participation but were given transportation and travel money at the conclusion of each scheduled visit. The respondent claimed financial compensation would be of little concern to him; if something were to happen, it would be treatment that was needed, not money. However, he believed participants should not expect ancillary and/or post-trial care if they agreed it was not part of the initial consent agreement. After he noted the great risk participants assumed by being tested with experimental drugs, he added that it would be appreciated if investigators decided to provide these services on their own, despite the consent agreement.

A few respondents also mentioned interest in gaining general knowledge about clinical research and its processes as a driving factor supporting involvement in the trial. A 24-year-old male mathematics and economics student further elaborated by explaining:

*I'm interested in those kinds of research and being a student myself ... I like to learn from different areas. And I feel it's like an opportunity to learn something, so yeah. Take part and then you learn what it's about, so it's also sometimes widen your horizons, so it all helps us.*

A number of individuals voiced concern regarding situations in which trials were not primarily intended to benefit the local communities and questioned what investigators hoped to gain from this activity:

*I think that I want to know who they are and where are they coming from and why are they doing it, and why you doing what you are doing. Why are you doing it?! And once I'm sure of who you are and sure [then maybe it is ok to participate]* (26-year-old male, information and Spanish student)

Respondents' level of education did not give evidence to any differences in attitudes on clinical trial involvement. The 30-year-old high school graduate

included in this study, one of the few who did not have a university-level education, also regarded clinical trials as worthwhile to the public. He suggested
that conducting clinical trials in the country could also lead to the production of higher quality drugs that are better suited to African communities.

Two of the twenty respondents mentioned some type of compensation
as being primarily responsible for generating initial interest in the trial. One
respondent admitted to "hoping to receive any gift" through participation,
whether it be a gift bag or some other item. In the other case, an account was
given of how the search for a summer job had mistakenly led to interpreting
the trial poster as offering an internship opportunity:

> To be frank, initially, when I saw the notice I was kind of looking for vacation
> internship. So I was on vacation and I came to school looking for internship and I
> saw the notice and thought I was an opportunity. So I just wanted to give it a try
> thinking it was something that I was going to benefit financially, that was the first
> thing that got me. But after coming and realizing that it was a clinical trial and I
> wasn't coming to be an intern—and get some experience and some small little thing
> in my pocket—and well I just said let me just be a part of it. It's also an experience,
> so that's why I joined. (25-year old male, psychology student)

After realizing the trial was not a job opportunity, he became interested
in learning about the research process itself. He stated that the opportunity
to be part of an actual clinical study and see how it is conducted became his
main motivation.

## Respondents' Views on Trial Purposes

The idea that the phase I vaccine trial was ultimately meant to benefit citizens of Ghana and Africa alike was shared by a number of respondents. The
respondent who earlier made mention of positioning the trial as a "study"
continued to give a more contextual overview of what investigators were hoping to achieve by saying:

> They were trying to look at a vaccine that was going to help controlling malaria
> which is a serious problem in Africa and I felt like it was something worth support-
> ing. (26-year-old male, environmental science student)

Over the course of the study, respondents continued to emphasize their
belief that the trial was meant to address Ghana's malaria problem. This belief
appeared to be grounded in commonly held values that "medical" research
and the investigators conducting it were directing their efforts toward the

betterment of society. Referring to the question of what investigators hoped to achieve through the trial, a respondent asserted:

*What they say they were going to use it for, they should use it for. It is a new vaccine they are going to use it to help the poor, the needy, the rich, everyone. The child, the old, everyone in this country so (proceeds to emphatically claps hands) … so that is what they were saying. They would use it to help the country!* (26-year-old male, information and Spanish student)

Respondents supported the idea that locally conducted research should benefit those in the community and, to a broader extent, the country. However, it was made clear this did not mean that the benefits produced through research should only be restricted to Ghanaians or Africans. A 31-year-old male psychology student having participated in the trial stated:

*It's not just limited … you shouldn't get me wrong. The priority is for my people but it's not that it should be limited to my people, no, no. It can go out, and once it is very effective here, other people can come and use it. Even the [pharmaceutical] producers will see once it's doing good here, in this area and within Ghana and Nigeria … we are within the west continent and we have a very lot in common. It will be very difficult for you to distinguish between a Nigerian and a Ghanaian. So once it's doing well here, there are chances it can also do well there. And even within Africa and beyond. But the main priority is here, from here than we look outside.*

A number of individuals offered vague details when recalling information that had been given about the vaccine's development and the trial purpose. There was a definite sense of uncertainty expressed in these respondents' voices. When asked if they were generally informed or explicitly told about the purpose of the study, a respondent explained:

*Yes, I think I was but I forgotten a little. I think they told us that they've tested a vaccine in some other parts of the [world], now they are coming to test it on Africans to see whether it will also fit … I think … I know that vaccination its prevention, so that was all I know.* (21-year-old female, English and political science student)

This uncertainty was also observed in another interview with a respondent who participated in the trial:

*But we were not told whether we we you were on the vaccine or not. And we were told what makes up the vaccine … I think it was on the document that we all took. But for me, I didn't really consider so much what makes up the vaccine.*

*So some vaccines, some HP (hepatitis) … these things, we don't know what it is, what they are talking about.* (26-year-old male, information and Spanish student)

The vague descriptions of trial protocols seemed to indicate that individuals were not concerned with trial particulars. Furthermore, respondents expressed that trial participants had an equal role and voice in the research process. They believed that since investigators had the same goal in mind, there was less of a need to be concerned with the "minor" details.

Some respondents were convinced the trial was not interested in investigating reactions or the vaccine's safety, but rather its efficiency. One respondent who took part in the trial also felt that vaccine testing was being done to see "whether it works" and "how it works." Many respondents could not imagine that testing for toxicity or safety would be done on them, with some abruptly dismissing the possibility:

*They were not testing how much is safe but they were testing its effectiveness. Because I believe it's malaria, and malaria is pronounced here in Africa. And they did test it in other parts of the world. And Africa also needs it! And they have to test it here! And it was necessary for them to do it here so that … (grunt). Ask me the next question.* (25-year-old male, psychology student)

Overall, respondents concluded that, based on what they were told by investigators during the information and consent meeting, participation in the vaccine trial would bring them no significant harm, although there was differing opinions as to whether investigators were examining the vaccine's efficacy or safety.

## Comprehending Biomedical "Jargon"

As mentioned earlier, many respondents had difficulty describing the meaning of the research phases. Others, while not being familiar with the terminology, were able to accurately describe activities taking place within the trial that corresponded with phase I research. They correctly explained that the trial was using both an experimental vaccine, which might not work, as well as a placebo control. Those able to accurately articulate these activities were typically respondents who mentioned an interest in being educated about clinical trial processes and requested to be updated on trial developments after their participation responsibilities had concluded. A trial participant revealed how he became educated on trial methodology and the use of placebos through his psychology studies:

*Well, at the university I read about psychology and in psychology there's these instances where you can volunteer to participate in a research course. They even mention instances where you can even go in and you will not be even injected with the vaccine itself. You will either be given a placebo or something. So I was enlightened from the university.* (31-year-old male, psychology student)

Other ideas regarding placebo use were varied among respondents. Some were completely unaware of placebos; others believed that they were just a weaker form of the vaccine. A small number of respondents had a firmer understanding of placebo use in research. One of the respondents who constantly expressed a keen interest in being educated on trial activities at all stages of research and postparticipation stated:

*That's what they were doing. They had people that were given the drug and others who were given the salt solution. So, I learned that's what they were doing so they had a control experiment or something like that. But they never revealed which people were taking the placebo or which people were taking the vaccine.* (20-year-old male, economics student)

On the other hand, some respondents who participated in the trial and who had no prior knowledge of trial processes demonstrated that, after being informed about trial processes during the initial information session as well as participating in the trial itself, they were still unsure of the protocols used:

*I was told that there was some … I'm not sure this is right … there was some plasma something … they were putting into our bloodstream or something to check its reaction.* (26-year-old male, information studies and Spanish student)

It was initially unclear what was meant by "check its reaction." Further questioning aimed to clarify if he was aware that investigators wanted to discover if the vaccine would cause adverse reactions, or whether he assumed they wanted to confirm the effectiveness. It became apparent that he was referring to the former. This was reinforced by his description of the tools he was given to observe and document reactions, stating that they were given a thermometer, a notebook, and a ruler to measure swelling at the injection site. He was also able to recall that investigators had told him this research had been conducted abroad. He believed it was previously conducted in America; now investigators wanted to test it in Africa to see whether it "fit" the local population. He referenced the informed consent sheet he and other respondents were given to back his claim that Africans have different "bodies" than Americans and Europeans. As he suggested, this was derived, in part, from living in different environmental conditions.

Others also acknowledged the significance of investigators' need to test the vaccine in different populations and settings. Respondents stated that the development of clinical interventions meant to be distributed in Ghanaian society should at some point in the research process include Ghanaians. This was stated as a measure to ensure the intervention developed was adaptable to Ghanaians. Correspondingly, it was believed that the work was being carried out in Ghana because it was necessary to see whether the environment, the climate, and local factors would play a part in the vaccine's effectiveness:

*Maybe drugs are made to cure a certain illness, but due to weather, climate, other environmental factors a drug that is effective in North America may not be effective in Ghana here, or may not be 'relatively' effective in Ghana. So if me, in this climate, are being used in this clinical trial, I think the outcome of it will suit the environment and the people here more than those in North America. (24-year-old male, mathematics and economics student)*

The response reflected the tendency of respondents to embed answers about trial information, concepts, and processes in the context of what the trial meant for Ghana and the Ghanaian people.

## Discussion and Conclusion

Overall, the present study found that the meaning and value respondents constructed around various research concepts (e.g., informed consent) were often grounded in beliefs that research aimed to improve local health issues. Values attributed to participation were developed within a wider context of influences that originated outside of the informed consent process. These influences included considerations about the local health situation in which malaria was considered a significant threat, coupled with individuals' desire to contribute to positive research conducted by investigators they trusted. This corresponds to the work of Felt et al. (2009), who argue that individuals' ways of understanding and knowing occur in a much wider context than that found or afforded by informed consent alone. Furthermore, the respondents, who generally considered participation in research as a way to contribute to the improvement of local health, characterized the nature of the work being done as a public good, having high social value. Such framing was also seen to play a role in respondents' decision making, risk consideration, and research expectations. While the results of this study are not meant to be a broadly generalized reflection of other research populations and their perceptions of research and participation, this study provides the opportunity to benefit from insights offered by a particular group of individuals who have expressed interest and/or participation in clinical research. We also acknowledge that respondents of a Phase I trial may provide different perspectives of research participation from those participating in later stage research. Phase I clinical trials often come with more risk than late-phase trials, as Phase I research involves the initial application of experimental drugs on humans with no guarantee of effectiveness (Nurgat et al. 2005; Shamoo and Resnik 2006).

The limitations of informed consent in relation to local influences have been questioned in several studies that examined the effectiveness of informed consent in low-income countries (Krosin et al. 2006; Mystakidou and Panagiotou 2009; Shapiro and Meslin 2001; Tindana, Kass, and Akweongo 2006). In a 2008 study investigating informed consent among Ghanaian women in a placebo-controlled supplementation trial, Hill et al. (2008) concluded that the traditional approach to informed consent was not adequate for conveying knowledge of trial protocols and activities. They further claimed that relational dynamics were influential to participants' understandings of trial related information. Similar findings regarding relational influences were found in studies conducted in other low-income (and emerging) countries including South Africa, Bangladesh, Gambia, and Uganda (Fitzgerald et al. 2002; Joubert et al. 2003; Leach et al. 1999; Lynoe et al. 2001). It would be naive to assume the complexities of informed consent that have led to arguments about its adequacy and use are issues limited to low-income settings. Joffe et al. (2001), investigating the quality of informed consent in cancer clinical trials in the United States, also highlight that there is a need to critically reflect on the appropriateness and straightforward application of informed consent in diverse research settings. In this study, the utility of informed consent, as it is generally understood, was called into question as respondents seemed to express understandings of research concepts contained within informed consent that were derived from influences beyond the information paradigm embedded in the informed consent process.

Further, studies exploring informed consent in biomedicine in the African setting have supported the idea that individuals hailing from poorer African communities make easy recruiting targets for vaccine trials; they tend to be from low socioeconomic groups, illiterate, and desperate for help (Frimpong-Mansoh 2008), also making them more susceptible to therapeutic misconception (TM). Some literature, which discusses the extent of what is considered to be TM, would categorize the respondents in the present study as "vulnerable and susceptible to exploitation" (Pentz et al. 2012) because they did not "seem" to understand that the research tested safety and not effectiveness (Appelbaum et al. 2012; Silverman 2011). However, the majority of respondents seeking out clinical trial participation in this study were university-educated students and did not express any signs of feeling exploited or being a part of a vulnerable group.

The need to supply prospective participants with information to enable thoughtful and responsible decision making is evident. However, there remains debate as to what that information should be, how it could be communicated, and how much context should play a role in determining legitimate understandings. It seems too simple to assume the ways in which information is communicated, received, and understood can be expected to be confined to a standardized informed consent procedure. Subsequently, calls for strengthened protection mechanisms that support increasing standardization and implementation of a decontextualized, objective informed consent procedure might actually undermine the intended outcome. Practices that predefine what is considered to be important information and methods of communication tend to overlook alternative ways of knowing; these practices also discourage exploration of the types of information participants deem relevant and the methods through which this occurs. The practices bring attention to the lack of influence and inclusion afforded to participants and, more broadly, the general public in matters related to research governance frameworks and what are considered to be ethical practices of research involving humans. Traditionally, the dominant voice in biomedical research governance discussions has been reserved for professional "experts" and academics. There is a need to strive for greater public inclusion in discussions by affording the public and research participants a voice in current debates in order to improve current research frameworks. Inclusion requires greater democratization in biomedicine that allows greater influence and appreciation for alternative ways of knowing and relational decision making. Allowing diverse stakeholders, including participants, a voice in research discussions serves as an important ethical instrument; it provides a democratic space in which members of society and citizens who engage in biomedical research attempts to bring their own feelings, values, beliefs, traditions, behaviors, and morality into the development and governance of technology and science.

## Notes

1. Sponsored by the Tanzanian Commission for Science and Technology
2. There are four clinical trial phases of drug development: Phase I—screening for safety among small group; Phase 2—efficacy and further safety testing; Phase 3—confirming effectiveness, monitoring side effects; Phase 4—post-marketing surveillance trials.
3. Failure to participate was either due to medical reasons or school related commitments.

# References

Angell, M. 1997. "The Ethics of Clinical Research in the Third World." *The New England Journal of Medicine* 337(12): 847–849.

Appelbaum, P. S., M. Anatchkova, K. Albert, L. B. Dunn, and C. W. Lidz. 2012. "Therapeutic Misconception in Research Subjects: Development and Validation of a Measure." *Clinical Trials (London, England)* 9(6): 748–61. Accessed July 7, 2013. http://www.ncbi.nlm.nih.gov/pubmed/22942217.

Appelbaum, P. S., L. H. Roth, and C. Lidz. 1982. "The Therapeutic Misconception: Informed Consent in Psychiatric Research." *International Journal of Law and Psychiatry* 5(3–4): 319–29. Accessed July 7, 2013. http://www.ncbi.nlm.nih.gov/pubmed/6135666.

Appelbaum, P. S., L. H. Roth, C. W. Lidz, P. Benson, and W. Winslade. 1987. "False Hopes and Best Data: Consent to Research and the Therapeutic Misconception." *Hastings Center Report* 2: 20–24.

Ballantyne, A. J. 2010. "How to Do Research Fairly in an Unjust World." *The American Journal of Bioethics:* 10(6): 26–35. Accessed July 7, 2013. http://www.ncbi.nlm.nih.gov/pubmed/20526966.

Beauchamp, T., and J. Childress. 2009. *Principles of Biomedical Ethics.* Sixth Edition. New York, New York: Oxford University Press.

Belmont Report. 1979. *The Belmont Report: Ethical Principles and Guidelines for the Protection of Human Subjects of Research.* Washington, DC. Accessed June 12, 2013. hhs.gov/ohrp/humansubjects/guidance/belmont.html.

Benatar, S. R. 2002. "Reflections and Recommendations on Research Ethics in Developing Countries." *Social Science and Medicine (1982)* 54(7): 1131–1141. Accessed July 7, 2013. http://www.ncbi.nlm.nih.gov/pubmed/11999507.

Brody, B. A. 2002. "Ethical Issues in Clinical Trials in Developing Countries." *Statistics in Medicine* 21(19): 2853–2858.

Buchanan, D., S. Sifunda, N. Naidoo, S. James, and P. Reddy. 2008. "Assuring Adequate Protections in International Health Research: A Principled Justification and Practical Recommendations for the Role of Community Oversight." *Public Health Ethics* 1(3): 1–12.

Chingono, A., T. Lane, A. Chitumba, M. Kulich, and S. Morin. 2008. "Balancing Science and Community Concerns in Resource-Limited Settings: Project Accept in Rural Zimbabwe." *Clinical Trials (London, England)* 5(3): 273–6. Accessed July 7, 2013. http://www.ncbi.nlm.nih.gov/pubmed/18559417.

Corrigan, O. 2003. "Empty Ethics: The Problem with Informed Consent." *Sociology of Health and Illness* 25(3): 768–792.

Council for International Organizations of Medical Sciences (CIOMS). 2002. *International Ethical Guidelines for Biomedical Research Involving Human Subjects.* Geneva, Switzerland.

Dixon-Woods, M., S. J. Williams, C. J. Jackson, A. Akkad, S. Kenyon, and M. Habiba. 2006. "Why Do Women Consent to Surgery, Even When They Do Not Want To? An Interactionist and Bourdieusian Analysis." *Social Science and Medicine (1982)* 62(11): 2742–2753.

Dunn, M., M. Sheehan, T. Hope, and M. Parker. 2012. "Toward Methodological Innovation in Empirical Ethics Research." *Cambridge Quarterly of Healthcare Ethics* 21(4): 466–480. doi:10.1017/S0963180112000242.

Ehrich, K., C. Williams, B. Farsides, J. Sandall, and R. Scott. 2007. "Choosing Embryos: Ethical Complexity and Relational Autonomy in Staff Accounts of PGD." *Sociology of Health and Illness* 29(7): 1091–1106.

Ekouevi, D. K., R. Becquet, I. Viho, L. Bequet, C. Amani-Bosse, F. Dabis, and V. Leroy. 2004. "Obtaining Informed Consent from HIV-Infected Pregnant Women, Abidjan, Cote d' Ivoire." *AIDS* 18(10): 1486–1488.

Ellis, R. D., I. Sagara, A. Durbin, A. Dicko, D. Shaffer, L. Miller, J. Millum, et al. (2010). "Comparing the Understanding of Subjects Receiving a Candidate Malaria Caccine in the United States and Mali." *The American Journal of Tropical Medicine and Hygiene* 83(4): 868–872.

Emanuel, E. J., D. Wendler, J. Killen, and C. Grady. 2004. "What Makes Clinical Research in Developing Countries Ethical? The Benchmarks of Ethical Research." *The Journal of Infectious Diseases* 189(5): 930–937. doi:10.1086/381709.

Faden, R., T. Beauchamp, and N. King. 1986. *A History and Theory of Informed Consent.* Oxford: Oxford University Press.

Felt, U., M. D. Bister, M. Strassnig, and U. Wagner. 2009. "Refusing the Information Paradigm: Informed Consent, Medical Research, and Patient Participation." *Health (London, England: 1997)* 13(1): 87–106. doi:10.1177/1363459308097362.

Fitzgerald, D. D. W. D., C. Marotte, R. R. I. Verdier, W. D. J. Johnson, and J. W. Pape. 2002. "Comprehension during Informed Consent in a Less-Developed Country." *Lancet* 360(9342): 1301–1302. Accessed July 7, 2013. http://www.sciencedirect.com/science/article/pii/S0140673602113389.

Food and Drug. 2012. "Food and Drugs Board: Current On-Going Clinical Trials." *Food and Drugs Authority, Ghana.* Accessed June 15, 2012. http://www.fdaghana.gov.gh/pdfs/Quick links/ON-GOING CLINICAL TRIALS1.pdf.

Frimpong-Mansoh, A. 2008. "Culture and Voluntary Informed Consent in African Health Care Systems." *Developing World Bioethics* 8(2): 104–114.

Gatter, R. 2006. "Conflicts of Interest in International Human Drug Research and the Insufficiency of International Protections." *Am. J.L. and Med.* 32(2): 351–364. Accessed July 07 2013. http://heinonlinebackup.com/hol-cgibin/get_pdf.cgi?handle=hein.journals/amlmed32&section=21.

Ghana-Michigan Collaborative. 2010. "The Ghana-Michigan Collaborative Health Alliance for Reshaping Training, Education, and Research (Charter)."

Glickman, S. W., J. G. McHutchison, E. D. Peterson, C. B. Cairns, R. A. Harrington, R. M. Califf, and K. A. Schulman. 2009. "Ethical and Scientific Implications of the Globalization of Clinical Research." *The New England Journal of Medicine* 360(8): 816–823. Accessed July 7, 2013. http://www.ncbi.nlm.nih.gov/pubmed/19228627.

Goldberg, D. S. 2011. "Eschewing Definitions of the Therapeutic Misconception: A Family Resemblance Analysis." *The Journal of Medicine and Philosophy* 36(3): 296–320.

Hill, Z., C. Tawiah-Agyemang, S. Odei-Danso, and B. Kirkwood. 2008. "Informed Consent in Ghana: What Do Participants Really Understand?" *Journal of Medical Ethics* 34(1): 48–53.

Horng, S., and C. Grady. 2003. "Misunderstanding in Clinical Research: Distinguishing Therapeutic Misconception, Therapeutic Misestimation, and Therapeutic Optimism." *IRB: Ethics and Human Research* 25(1): 11–16.

Joffe, S., E. F. Cook, D. Cleary, J. W. Clark, and C. Jane. 2001. "Quality of Informed Consent: A New Measure of Research Subjects." *J Natl Cancer Inst* 93(2): 139–147.

Joubert, G., H. Steinberg, E. van der Ryst, and P. Chikobvu. 2003. "Consent for Participation in the Bloemfontein Vitamin A Trial: How Informed and Voluntary?" *American Journal of Public Health* 93(4): 582–584.

Klitzman, R. L. 2012. "US IRBs Confronting Research in the Developing World." *Bioethics* 12(2): 1471–8731. doi:10.1186/1472-6939-12-13.16.

Krogstad, D. J., S. Diop, A. Diallo, F. Mzayek, J. Keating, O. A. Koita, and Y. T. Touré. 2010. "Informed Consent in International Research: The Rationale for Different Approaches." *The American Journal of Tropical Medicine and Hygiene* 83(4): 743–747. Accessed July 7, 2013. http://www.pubmedcentral.nih.gov/articlerender.fcgi?artid=2946735&tool=pmcentrez&rendertype=abstract.

Krosin, M. T., R. Klitzman, B. Levin, J. Cheng, and M. L. Ranney. 2006. "Problems in Comprehension of Informed Consent in Rural and Peri-Urban Mali, West Africa." *Clinical Trials* 3(3): 306–313.

Kuczewski, M. G., and P. Marshall. 2002. "The Decision Dynamics of Clinical Research: The Context and Process of Informed Consent." *Medical Care* 40(9 Suppl): V45–54.

Leach, A., S. Hilton, B. M. Greenwood, E. Manneh, B. Dibba, A. Wilkins, and E. K. Mulholland. 1999. "An Evaluation of the Informed Consent Procedure Used during a Trial of a Haemophilus Inuenzae Type B Conjugate Vaccine undertaken in The Gambia, West Africa." *Social Science and Medicine* 48: 139–148.

Levitt, M., and H. Zwart. 2009. "Bioethics: An Export Product? Reflections on Hands-On Involvement in Exploring the "External" Validity of International Bioethical Declarations." *Journal of Bioethical Inquiry* 6(3): 367–377.

Lock, M., and V. Nguyen. 2010. *An Anthropology of Biomedicine.* Oxford: Wiley-Blackwell.

London, L., A. Kagee, K. Moodley, and L. Swartz. 2012. "Ethics, Human Rights and HIV Vaccine Rrials in Low-Income Settings." *Journal of Medical Ethics* 38(5): 286–293. doi:10.1136/medethics-2011-100227.

Lorenzo, C., V. Garrafa, J. H. Solbakk, and S. Vidal. 2010. "Hidden Risks Associated with Clinical Trials in Developing Countries." *Journal of Medical Ethics* 36(2): 111–5.

Lynoe, N., Z. Hyder, M. Chowdhury, and L. Ekstrom. 2001. "Obtaining Informed Consent in Bangladesh." *New England Journal of Medicine* 344(6): 460–461.

Mansell, P. 2011. "Over 50% Growth to 2015 Seen in Global Trials Market." *PharmaTimes.* Accessed August 5, 2012. http://www.pharmatimes.com/article/11-07-07/Over_50_growth_to_2015_seen_in_global_clinical_trials_market.aspx.

Marshall, P. A., C. A. Adebamowo, A. A. Adeyemo, T. O. Ogundiran, M. Vekich, T. Strenski, J. Zhou, T. E. Prewitt, R. S. Cooper and C. N. Rotimi 2006. "Voluntary

Participation and Informed Consent to International Genetic Research." *American Journal of Public Health* 96(11): 1989–1995.

Mbuagbaw, L., L. Thabane, P. Ongolo-Zogo, and T. Lang. 2011. "The Challenges and Opportunities of Conducting a Clinical Trial in a Low Resource Setting: The Case of the Cameroon Mobile Phone SMS (CAMPS) Trial, An Investigator Initiated Trial." *Trials* 12(1): 145.

Miller, T., and M. Boulton. 2007. "Changing Constructions of Informed Consent: Qualitative Research and Complex Social Worlds." *Social Science and Medicine* 65: 2199–2211. Accessed July 7, 2013. http://www.sciencedirect.com/science/article/pii/S0277953607004340.

Moreno, J. D. 2007. "Goodbye to All That: The End to Moderate Protectionism in Human Subjects Research." *The Hastings Center Report* 31(3): 9–17.

Morris, N., and B. Bàlmer. 2006. "Volunteer Human Subjects' Understandings of Their Participation in a Biomedical Research Experiment." *Social Science and Medicine (1982)* 62(4): 998–1008. doi:10.1016/j.socscimed.2005.06.044.

Moses, H., R. Dorsey, D. H. M. Matheson, and S. O. Thier. 2005. "Financial Anatomy of Biomedical Research." *JAMA* 294(11): 1333–1342.

Mystakidou, K., and I. Panagiotou. 2009. "Ethical and Practical Challenges in Implementing Informed Consent in HIV/AIDS Clinical Trials in Developing or Resource-Limited Countries." *Journal of Social Aspects of HIV/AIDS* 6(2): 37–41.

Nurgat, Z., W. Craig, N. Campbell, J. Bissett, J. Cassidy, and M. Nicolson. 2005. "Patient Motivations Surrounding Participation in Phase I and Phase II Clinical Trials of Cancer Chemotherapy." *British Journal of Cancer* 92(6): 1001–1005. doi:10.1038/sj.bjc.6602423.

Oduro, A. R., R. A. Aborigo, D. Amugsi, F. Anto, T. Anyorigiya, F. Atuguba, A. Hodgson and K. A. Koram, et al. 2008. "Understanding and Retention of the Informed Consent Process among Parents in Rural Northern Ghana." *BMC Medical Ethics* 9(12): 12. Accessed August 7, 2013. http://www.pubmedcentral.nih.gov/articlerender.fcgi?artid=2443367&tool=pmcentrez&rendertype=abstract.

Ogutu, B., R. Baiden, D. Diallo, P. Smith, and F. Binka. 2010. "Case Study Sustainable Development of a GCP-Compliant Clinical Trials Platform in Africa: The Malaria Clinical Trials Alliance Perspective." *Malaria Journal* 9(103): 1–10. Accessed July 7, 2013. http://www.biomedcentral.com/content/pdf/1475-2875-9-103.pdf.

Pentz, R. D., M. White, R. D. Harvey, Z. L. Farmer, Y. Liu, C. Lewis, O. Dashevskaya, T. Owonikoko and F. R. Khuri. 2012. "Therapeutic Misconception, Misestimation, and Optimism in Participants Enrolled in Phase 1 Trials. *Cancer* 118(18): 4571–4578. Accessed August 5, 2013. http://www.ncbi.nlm.nih.gov/pubmed/22294385.

Personal Communication. 2012.

Petryna, A. 2007. "Clinical Trials Offshored: On Private Sector Science and Public Health." *BioSocieties* 2(1): 21–40. Accessed July 7, 2013. http://www.palgrave-journals.com/doifinder/10.1017/S1745855207005030.

Petryna, A. 2009. *When Experiments Travel: Clinical Trials and the Global Search for Human Subjects*. Princeton: Princeton University Press.

Resnik, D. B. 2009. "The Clinical Investigator-Subject Relationship: a Contextual Approach." *Philosophy, Ethics, and Humanities in Medicine:* 4: 16. Accessed July 7, 2013. http://www.pubmedcentral.nih.gov/articlerender.fcgi?artid=2794289&tool=pmcentrez&rendertype=abstract.

Reynolds, W. W., and R. M. Nelson. 2007. "Risk Perception and Decision Processes Underlying Informed Consent to Research Participation." *Social Science and Medicine (1982)* 65(10): 2105–15. doi:10.1016/j.socscimed.2007.06.021.

Sariola, S., and B. Simpson. 2011. "Theorising the 'Human Subject' in Biomedical Research: International Clinical Trials and Bioethics Discourses in Contemporary Sri Lanka." *Social Science and Medicine* 73(4): 515–521.

Schüklenk, U. 2010. "For-Profit Clinical Trials in Developing Countries—Those Troublesome Patient Benefits." *The American Journal of Bioethics, 10*(6), 52–54.

Shamoo, A. E., and D. B. Resnik. 2006. "Strategies to Minimize Risks and Exploitation in Phase One Trials on Healthy Subjects." *The American Journal of Bioethics:* 6(3): W1–13.

Shapiro, H. T., and E. M. Meslin. 2001. "Ethical Issues In the Design and Conduct of Clinical Trials in Developing Countries." *N Engl J Med* 345(2): 139–142.

Silverman, H. 2011. "Protecting Vulnerable Research Subjects in Critical Care Rrials: Enhancing the Informed Consent Process and Recommendations for Safeguards." *Annals of Intensive Care* 1(1): 8. doi:10.1186/2110-5820-1-8.

Stoljar, N. 2011. "Informed Consent and Relational Conceptions of Autonomy." *The Journal of Medicine and Philosophy* 36(4): 375–384.

Sugarman, J., N. E. Kass, S. N. Goodman, P. Perentesis, R. R. Faden, and P. Fernandes. 2012. "What Patients Say about Medical Research." *Ethics* 20(4): 1–7.

Terranova, G., M. Ferro, C. Carpeggiani, V. Recchia, L. Braga, R. C. Semelka, and E. Picano. 2012. "Low Quality and Lack of Clarity of Current Informed Consent Forms in Cardiology: How to Improve Them." *JACC. Cardiovascular Imaging* 5(6): 649–655.

Tindana, P. O., N. Kass, and P. Akweongo. 2006. "The Informed Consent Process in a Rural African Setting: A Case Study of the Kassena-Nankana District of Northern Ghana." *IRB* 28(3): 1–6.

Van Loon, K., and G. Lindegger. 2009. "Informed Consent in Clinical Trials: Perceptions and Experiences of a Sample of South African Researchers." *Health SA Gesondheid* 14(1): 1–7.

Zvonareva, O., N. Engel, E. E. Ross, R. Berghmans, A. A. A. Dhai, and A. Krumeich. 2013. "Engaging Diverse Social and Cultural Worlds: Perspectives on Benefits in International Clinical Research From South African Communities." *Developing World Bioethics* 8731. doi:10.1111/dewb.12030.

# CHAPTER 2

# Local Perspectives on Universal Bioethics—A Qualitative Study on Informed Consent in South India

*Inge A. S. van Alphen, Nora Engel, and Mario Vaz*

## Introduction

Increasingly, pharmaceutical phase III clinical trials are coordinated across multiple centers and clinical research organizations worldwide (Annas 2009). In line with this global development, international cooperation has universalized global health standards and has standardized medical bioethics. For the sake of patient protection worldwide, the Declaration of Helsinki was formulated by the World Medical Association during the 1960s. It upholds internationally-recognized principles, guidelines, and norms of medical ethics that serve to protect human subjects during medical research; it also stipulates the three linchpins of high-quality informed consent: the forbidding of coercion, the mandate that patients who volunteer fully comprehend the risks associated with the research in question, and the stipulation that consent must be formally documented and witnessed (Lee 2010).

However, concerns about participant rights in clinical trials—particularly regarding the process of how informed consent is obtained from participants—have been increasingly voiced (Fitzgerald et al. 2002; Shah 2006; Strauss et al. 2001; Rajaraman et al. 2011; Silverman 2011). In essence, there are concerns that governance of patient protection during clinical trials is ineffective, especially in low- or middle-income countries. Qualitative studies on clinical trial participation reveal that cultural and contextual factors affect the degree to which international standards are applicable in a local setting. Examples of confusion about study terminology or incomplete disclosure have been documented, and the enrollment of participants who are ill-informed about their choices has also been witnessed. This has led to the realization among scholars that the standardization of bioethics may not result in true informed consent (Strauss et al. 2001; Shah 2006; Rachid 2006; Lee 2010; Miller 2002; Maiti 2007).

Locally, the nature of the potential participants can pose daunting challenges for the procurement of informed consent. No matter how stringent or thorough international standards of bioethics may be, the reality is that patient understanding is framed by norms, language, society, intellect, culture, and education. These barriers are context-specific; the achievement of true informed consent, in such cases, relies on "translating" the global standards to a local setting by using terminology and metaphors that are familiar to the local population. Several studies, such as the one by Sengupta and his colleagues (2011), underline the fact that universalizing standard procedures may be an example of a swinging door. It may, in fact, hamper patient protection because international standards are not responsive enough to local customs if they are stringently implemented. For example, international standards require lengthy and difficult terminology on consent forms. As a result, the quality of the informed consent process is often lost during the process of translating it to local settings; extra effort and innovation are applied in order to ensure that patients fully comprehend the details of the research study in which they are enrolling. The adjustment of the information sheet for the informed consent procedure is an example of physicians aligning international standards to local needs. To quote Christian Byk (2009), a judge, general secretary of the International Association of Law, Ethics, and Science, and member of the French Commission for UNESCO: "What is the point of drawing up an international code of bioethics if we remain convinced that the diversity of cultures gives a different and even divergent meaning and scope to ethical principles?"

This chapter examines the perspectives of physicians, members of ethics boards, members of contract research organizations, and principle investigators in order to shed light on the extent to which the universalization of the process of obtaining informed consent has resulted in real, successful informed consent in terms of patient understanding. Semi-structured, qualitative interviews were undertaken in a South Indian private hospital and medical research center with actors involved in the conduct of clinical trials. These interviews asked participants to detail the attitudes and values surrounding the barriers to informed consent in a local setting. The insights from the interviews with principle investigators and physicians contribute to the discussion around the use of a universal, mainstream method. The analysis also highlights the sociocontextual challenges relevant to consent taking in a local setting and provides insights and suggestions to improve the quality of informed consent. The results shed light on how the process of informed consent can be made both more rigorous and more adaptive to local settings in a context-based manner.

## Standardizing Biomedical Research: Informed Consent

As a result of globalization, research and development has expanded globally. By outsourcing clinical trials to countries such as India, a country where costs of conducting trials are low, timely completion and rapid patient recruitment facilitate trial conduction. As such, pharmaceutical companies are able to put their drugs on the market sooner, reap a higher return on investments due to longer patent protection, and, consequently, yield higher profits (Laughton 2007). While only 2.7 percent of all clinical trials carried out globally between 2007 and 2011 were conducted in India, the number of trials conducted within the country increased during this period by 3.7 percent. At the same time, the numbers of registered trials in the United States and the EU decreased by 11 percent. As a result, Contract Research Organizations (CROs) have greatly increased in number. CROs perform clinical trials and subsequently compile and submit the results to industry and regulatory agencies on behalf of drug companies (Laughton 2007). Receiving countries have driven this process by promoting their pharmaceutical industries and the practice of clinical trials by foreign pharmaceutical companies (RNCOS 2007).

The debate about how to best govern and rule international medical research is ongoing. Informed consent is one of the cornerstones of medical research and has internationally recognized standards in the Declaration of Helsinki and the International Covenant on Civil and Political Rights (ICCPR). However, the legal status of the Declaration of Helsinki as part of international human rights law remains unclear (Annas 2009). While the guidelines in the declaration represent a minimum standard to protect research participants, they are operated through different national regulatory policies. This is problematic, as the governance that enforces human rights protection is inconsistent; the different national informed consent standards are presently uncoordinated and not aligned. There is no authoritative structure that monitors, for instance, the process of informed consent globally; there is no effective global court in which to challenge human rights abuses with regard to informed consent. Furthermore, there is a lack of understanding of how global standards are applied in local settings and what kind of practical barriers principal investigators (PIs) face; PIs are responsible for undertaking the research and observing the human rights, health, and welfare of the trial participants (ICMR 2006). Consequently, pharmaceutical companies and CROs are not held fully accountable for unethical behavior, such as involuntary recruitment, during clinical trials (Lee 2010). The lack of clear enforcement of codes of conduct, standard operating procedures, or good clinical practice (GCP), combined with the disregard of contextual factors of local trial settings such as India in the current regulatory policies, renders the degree of voluntariness of trial participation questionable.

On pace with the increasing influx of clinical trials in India, the demand for "safeguards to protect human participants subjected to biomedical research" grew (ICMR 2006, 6). In 2000, the Indian Council for Medical Research (ICMR) produced the Ethical Guidelines for Biomedical Research on Human Subjects. It stated, similarly to the Declaration of Helsinki, that all research proposals on biomedical research involving human participants must be cleared by an ethics committee that is entrusted with safeguarding the welfare and rights of participants. Under the auspice of the ethics committees, PIs must abide by international human rights standards. Frequently, these PIs are also clinicians in institutions such as governmental or private hospitals or clinics; subsequently, they have a large number of potential trial participants in their care.

However, the universality of the standard process is questioned when local patients with different cultural, religious, ethnic, educational, and social backgrounds are subjected to international standards. In low- and middle-income countries, such as India, where doctor-patient relations often need to be considered in a context of poverty and lack of access to health care, trial participants become easily vulnerable to inducement and misconception surrounding clinical trials. Furthermore, in settings with low levels of education and patient-health literacy and high levels of therapeutic misconception (when patients falsely believe that clinical trials have a therapeutic intent), patients often do not question their participation in clinical trials. In such a scenario, the concept of informed consent becomes extremely relevant (LaFraniere, Flaherty, and Stephens 2000).

Studies in Bangladesh (Zaman and Nahar 2011) have highlighted difficulties in conveying the meaning of the word "research," as the formal Bengali word amounts to "finding a lost cow." The same study reveals that some ethical codes for informed consent may appear absurd in Bangladeshi villages. Consequently, by challenging the universality of research ethics, the authors call for an "indigenization" of bioethics, or an adaptation of standardized codes of conducts. Similarly, Rajaraman and her colleagues (2011) conducted a study in which parents of newborn babies provided consent for children's participation in an observational tuberculosis research study. These parents and babies were followed for two years as part of the trial. The study highlighted that greater efforts need to be directed toward obtaining simplified communication materials for the process of informed consent. The authors emphasized the need to improve the communication skills of research workers with regard to "explaining research processes and putting potential research participants at ease" (Rajaraman et al. 2011, 1). Such insufficiencies in the process of informed consent become a real concern, especially for the poor, who have less access to medical treatment and therefore might seek it from clinical trials, or those who have strong trust-based relationships with the recruiting doctors (Williams 2012).

## Methodology

Sixteen semi-structured interviews were conducted with members of Contract Research Organization (CROs), the Institutional Ethics Review Board (IERB), physicians, principal investigators, coprincipal investigators, and ethicists in a private hospital and research center in South India. The hospital is run by a not-for-profit private society; the outpatient department (OPD) sees approximately 2,500–3,000 patients per day across the socioeconomic spectrum. The hospital also conducts medical research in the form of clinical trials for pharmaceutical companies, also through CROs. A large number of departments are involved in these trials, from pediatrics to psychiatry. These are phase III, randomized, multicenter trials that assess the therapeutic effect of new drugs in terms of safety and efficacy.

Interview guides were prepared in advance, based on literature review and analysis of research ethics guidelines. Respondents were asked to reflect on their experience with informed consent in clinical trials and to comment on patient perception of informed consent. Interview guides were framed around the interviewee's understanding, attitude, and values toward the process of obtaining informed consent in a local setting. Initial observations from the first two interviews were brought into the conversations of the remaining interviews, and the topic list was adjusted accordingly. The interview encounter was the context within which valid knowledge was produced; the emphasis was on conversation and interaction rather than question and answer. Therefore, arguments and debates were also part of the social interaction during the interview. The audio-recorded interviews were transcribed verbatim; preliminary analysis was done by progressive coding and categorization of issues that emerged from the interviews. The transcripts were read several times in order to identify the coding system. Major themes were identified through thematic analysis. The rich description of the data is useful because this research topic is one that is not well-known, and the participants' views on the topic are under-researched.

Interviewees were given a written information sheet, and they all signed informed consent forms. The study was approved by the IERB of the hospital and research center in question. All of the interview excerpts have been presented anonymously. For this reason, the details of the respondents are not highlighted, and the departments are not named. Presented below are the different perspectives and understandings that emerged from the interviews on the challenges in obtaining informed consent during clinical trials in a local setting. The results are structured in two parts: barriers to informed consent and different perspectives on solutions.

## Barriers to Informed Consent: Patient Understanding

Most respondents indicated that potential trial participants were recruited from the hospital outpatient department. A senior consultant of general medicine illustrates how recruitment can take place and that professionals may feel uncertain regarding the patient's understanding of the informed consent form:

> Here, patients are taken from the OPDs [outpatient departments], obviously, and generally the treating doctor asks his patients if they are interested, and then you know you try to tell them, some of them may not understand obviously, and you try to explain to them as best as possible and then you take the informed consent. But if they get what you are saying is very questionable. (Respondent 8, physician of general medicine, June 2012)

Strategies for recruitment vary between different categories of patients, and doctors expressed that patients carry a great deal of the responsibility when it comes to understanding the informed consent form or patient information sheet. Below are some of the challenges doctors face while taking informed consent.

### Education, Language, and Socioeconomic Status

Interviewees most commonly identified education, language, and socioeconomic status as the barriers to obtaining true and voluntary informed consent from patients and potential trial participants. A faculty member of the Board of Ethics emphasizes that education is an important barrier to obtaining informed consent. For example, she identifies that the level of education is directly linked to the amount of questions that a patient asks:

> An educated person will ask you more questions also. Not educated person will be explained and they will accept. Few people will say no I am scared. Especially if you talk to a lady alone, or something, she may be hesitant and may not participate. But

*in general, uneducated they don't ask.* (Respondent 9, professor and member of Institutional Ethics Review Board, June 2012)

An associate professor of medicine confirms another barrier related to educational status: a patient's education level contributes to possible vulnerability to enticement, because illiterate patients may be recruited without fully understanding:

> *I think the core issue is if somebody is well educated they may probably understand quicker or they may have much more questions in the beginning because they have already done their homework and come. They say look I have read up on the net that for schizophrenia these are the side effects and all that, so they have their knowledge. Somebody who is from a farm, they won't know much at all, and I will be the one giving them the information.* (Respondent 8, physician and associate professor of medicine, June 2012)

As expressed by this respondent, it is felt that a patient with low educational status will find it more difficult to understand the patient information sheet, especially if it is very long and complicated. Consequently, patients may rely on the trusting relationship with their doctors to obtain the information and ultimately make the decision on whether to participate. In India, there is no clear norm on the level of literacy for informed consent forms. Therefore, the importance of family members that are educated, or more knowledgeable, plays a crucial role.

Another factor identified is language; the language diversity between India's different states creates a barrier when the medical terminology on informed consent sheets consists of complicated legal nomenclature. A professor and faculty member in the Department of Medical Ethics mentions that the language is often too complicated for laymen patients:

> *Even in their local language, the pharmaceutical companies will use the terms (sic!) are very hi-fi, and the people will find it very difficult to understand. So even if it is a local language, it should be a little on the layperson's terms.* (Respondent 9, professor and faculty member in the Department of Medical Ethics, July 2012)

In addition to terminology itself, the translation of the informed consent sheets was also considered a barrier. A professor of psychiatry and member of the Institutional Ethics Review Board says:

> *Very often these consent forms are translated by professionals. The nuances of language they don't understand that. I don't understand my own mother tongue the way*

*it is professionally done because that's not the way in my daily life.* (Respondent 1, professor, physician, and Institutional Review Board member, June 2012)

Here, he identifies that the forms are not contextualized to the local population and setting and that this presents a problem for patients' understanding. Respondent 1 suggests that he is not in favor of professionally translated forms because they do not capture the colloquial language spoken daily by the participants. These forms are not applicable to the local setting, which will lead to a low level of patient understanding. Although the Drugs and Cosmetics Act of India specifies that the information should be "in a language that is nontechnical and understandable by the study subject" (Department of Health 2005, 4), this regulation does not ensure that patients are able to understand the terminology. The above quote by respondent 1 illustrates that the language used does not convey the essence or nuances of the regulation or a lack of harmony between national regulations and the needs of the local population. Due to this discord, the respondents feel that doctors and patients face barriers in achieving true and informed consent. Additionally, doctors must spend time providing verbal information to potential participants in order to ensure that they understand. The quality of the informed consent process may be hampered by the language barriers and the use of medical jargon on patient information sheets and informed consent forms. Most of the trials performed at the hospital and research center in question are drug trials for regulatory purposes, most of which are directly from multinational pharmaceutical companies or through CROs. These companies bring a ready-made informed consent form that is given to the PI, which needs to be approved by the respective Institutional Ethics Board. Doctors are also concerned with the length of the informed consent forms and patient information sheets, believing that this hampers obtaining true and voluntary informed consent during the recruitment process.

In addition to educational status, language, and literacy, the socioeconomic status of patients is another factor that respondents identified as a barrier to achieving true informed consent. Respondents voiced concerns that patients from a low socioeconomic situation may be more prone to inducement; consequently, the quality of informed consent is threatened since the voluntariness of the consent is questionable and may be a result of subtle coercion. Respondent 3 identifies free medication, less time spent in the outpatient department, and better treatment as reasons to consent to participation and feels that the direct benefits a patient may receive may act as an incentive:

*I'm saying right now the decision for me to sign is purely decided on: is my treatment free or not, am I getting beneficial, free, professional care, and more attention from you? He will get more attention, he will get priority. When I'm waiting for three hours in an outpatient, getting seen in the first 10 [minutes] is like giving me a bonus from heaven. I am not an idiot. I know if I sign and agree to your study I will get more attention. I will probably get a better drug.* (Respondent 3, a professor, physician, and faculty member of medical ethics, June 2012)

In this case, socioeconomic status is linked with the rewards of high quality care. Due to poverty and the lack of access to healthcare, respondents feel that poor Indian patients may participate in trials in order to receive medical attention or free medication. The link between socioeconomic status and level of incentive is displayed in the following quote; respondent 3 highlights that, depending on the socioeconomic status of the patient, the direct benefit to a patient may be something as trivial as receiving food or petrol:

*I may get some food to eat, some travel money. If I need food I am happy if you give a banana. If I have a car you give me a litre of petrol.* (Respondent 3, a professor, physician, and faculty member of medical ethics, June 2012)

He makes a direct link between financial status of the patient and incentive to participate in a trial. Paying patients to participate in clinical trials may be a form of undue enticement. In this case, the respondents are not talking about direct financial compensation for participation, but about receiving free medication and other benefits alongside trial participation.

Because many patients that come to the outpatient department are poor and come from a lower socioeconomic background, the respondents suggest that recruitment for trials may result in a lack of voluntariness and a degree of coercion. Coercion and voluntariness are two of the fundamental factors in the Declaration of Helsinki that must be avoided in order to obtain true and voluntary informed consent. Physicians articulated that informed consent is challenged by direct benefits from participating in a trial, whether it is free medication, less time spent in the outpatient department, or more attention from the doctors. Socioeconomic status and poverty level play a significant role in determining what constitutes a direct benefit.

### Patient-Doctor Rapport

One of the contextual factors specifically relevant to India is the doctor-patient relationship. Kumar (2008) identifies that, in India, doctors are held "in very high esteem" and that patients "typically proceed with treatment regimens as recommended by their practitioner" (Kumar 2008, 3). In addition to determining what constitutes a direct benefit to patients in financial or other forms, socioeconomic status also seems to play a role in the doctor-patient relationship and in the ease of patient recruitment. A good rapport between doctors and patients, especially poorer patients, is identified as a factor that may undermine the importance of true and voluntary informed consent. In such cases, patients trust their doctor with decision making, not assigning much importance to reading the long and complicated information sheets that are presented to them by their doctor. Although this is something which is culturally engrained in India (Kumar 2008), this factor seems to be more acutely present among patients of lower socioeconomic status. Doctors deal with patients from a wide range of socioeconomic backgrounds; respondent 4, a psychiatric physician and principle investigator, mentions that since poverty level is linked with the degree of trust that patients place in their doctor, the responsibility to ensure correct informed consent taking for the PI is increased:

> *In India the responsibility is higher, because the poorer the patient is the more they trust the doctor and the less likely they are to ask questions. Our more well-off populations as I have said I'm not worried because as I have said they will come and ask questions, they don't blindly accept, but for the others the doctor has to be much more responsible.* (Respondent 4, psychiatric physician and principle investigator, June 2012)

Respondents feel that socioeconomic status and level of education may determine the effort a patient undertakes to seek information from the doctor before agreeing to enroll in a trial. Generally, respondents expressed the feeling that the doctor-patient rapport inhibits true and voluntary informed consent because medical professionals are accustomed to making decisions on behalf of their patients. Respondent 5 emphasizes:

> *Culturally, at the level of the principal investigator, there is a feeling among many researchers that 'I know what is good for you, this is not going to be bad for you, why should I bother about reading you this whole thing?' At one extreme you may call it a paternalistic approach but on the other hand it's cultural. (…) Medical professionals are used to treating patients a certain way and sort of taking some decisions on their behalf. And now when they extend that same relationship, patient-doctor relationship, into researcher-subject relationship, then there may be a problem. That will be a cultural barrier.* (Respondent 5, physician and Institutional Ethics Review Board member, June 2012)

Patients from a lower socioeconomic class who have a lower educational status entrust the doctor with the decision of whether or not to enroll in a clinical trial. Many respondents expressed views that if the PI of the trial is also the patient's primary physician, this could represent an inducement. Respondent 3 expresses strong feelings about the unequal relationship between patients and physicians:

> You see the other issue is that especially in India the PI being the primary physician. That I think is the ultimate joke. If I am the guy asking you for consent, which normal guy would say no? That the primary physician who is looking after the patient is a PI that is the ultimate weapon, it's like a teacher asking me to participate. So I think that socioeconomic status is very important, (...) and then the physician-participant relation. (...) It is extremely unequal. (Respondent 3, professor, physician, and faculty member of medical ethics, June 2012)

Furthermore, respondents expressed that paying doctors in line with recruitment numbers is unethical; it represents a conflict of interest and could affect the quality of the informed consent. Hence, due to the nature of the relationship of the PI with the patient, the quality of the informed consent process may be jeopardized; the trial details and rights of the patients may be overlooked as a consequence of CROs hiring primary physicians as PIs.

However, some physicians also see it as an advantage to be a leading physician as well as PI; they can maximize the welfare of the patient as the drugs being tested in a trial are often beneficial, and patients also receive other benefits such as free tests. It was also articulated that the financial status of the patient can be a motivation to recruit patients into trials, especially those that are of low socioeconomic backgrounds and do not have access to expensive treatments. Respondent 12 says:

> See at least in places like here, when you look at the trial there are two three factors when you do a trial. One you find maybe a very good compound, a molecule, and it will benefit the patient because he clearly requires it and these medicines are very, very expensive. For example one medicine we give, we can converse in lakhs. One injection is about 1.5, 1.8 lakhs. It comes to almost 4000 US dollars, one injection. These are expensive, so you clearly want your sick patient to get it. (Respondent 12, professor of gastroentology, July 2012)

It is felt that there are both advantages and disadvantages for the leading physician to recruit patients to clinical trials. On one hand, doctors are supposed to have the welfare of their patients in mind; in addition to this, they have the responsibility to explain and make sure that each patient's informed consent process is conducted properly. However, the results also demonstrate that, due to the rapport between patients and their doctors, especially with patients from lower socioeconomic status, there is a certain sense of trust that

might undermine the need for the patient to ask questions or to inform him/
herself about the clinical trial in which they are enrolling.

## Improving Informed Consent Procedures

The respondents suggest that there are immediate and long-term needs
required to tackle the challenge of improving informed consent procedures
at a local level. The immediate needs should be reflected in recommendations that strengthen the quality of informed consent in relatively short time
periods. These recommendations are to increase the responsibility of Institutional Ethics Review Boards, to increase education and awareness for PIs,
and to evaluate patient understanding. The last recommendation involves
using local community engagement to understand the individual needs of
trial participants.

### *Strengthen the Role of Institutional Ethics Review Boards*

When respondents were asked about evaluating the process of informed consent, the majority pointed to the Institutional Ethics Review Board as a major
actor responsible for ensuring that PIs follow the ethical guidelines. They felt
that the role of the IERB goes beyond approving trials and includes supervision and monitoring. The respondents also felt that the IERB should monitor
informed consent, and research should be cleared only when there is enough
manpower to monitor the research. If the IERB cannot monitor the process
of informed consent, then it should not approve clinical trials. The following
quote by respondent 2 emphasizes this:

> *If the IRB [Institutional [Ethics] Review Board] cannot monitor, it should not allow
> so many studies to go through. I'm willing to accept, I think the IRB should under-
> stand its role. Its role is not to protect the institution; it's to protect the patients. The
> IRB keeps forgetting that its job is to protect the patient. Automatically the role is to
> monitor. Not enough manpower? Who is clearing the researches? Clear the research
> according to the manpower to monitor.* (Respondent 2, member of Institutional
> Ethics Review Board, June 2012)

Respondent 3 suggests that the Institutional Ethics Review Boards could
capitalize on the use of local knowledge to design context-specific patient
information and informed consent sheets:

> *You need to have ... when a PI sits down to write the informed consent, you need
> general physicians, maybe the IRB type of concept, we need a community person, we*

*need a physician, they should be sitting down to design the informed consent. That is where the IRB role should be.* (Respondent 3, professor, physician, and faculty member of medical ethics, June 2012)

Most of the respondents feel that the Institutional Ethics Review Board could be more involved in monitoring the informed consent process. Some suggestions were to spontaneously perform random checks; to video or audiotape the procedure (a practice which is now mandatory); and to uphold the ethical considerations that are required when administering informed consent. However, the respondents also identified that the review board is limited in its manpower; members are all volunteers and some work as full-time clinicians in the hospital or clinic. One of the recommendations, highlighted by the respondents, is to only allow clinical trials that the IERB is able to monitor. Another suggestion was that the institution cap the amount of clinical trials that can be run simultaneously. Although this was one of the themes identified throughout the interviews, respondents displayed two approaches to capping: cap the total amount of trials, or cap the number of trials run under one PI. Respondent 5, physician and present member of the Review Board in question, sheds light on the complexities involved when deciding what a numeric cap would entail. Depending on the size of the research team, the stage of the trial, and the personal involvement of the PI in the day-to-day activities of the trial, the workload will differ:

> *I can't think of a numeric cap, because it depends on the extent of his [the PIs] involvement. Now I'm a PI, and I have employed a whole research for this trial, I have another group there, then in that case my only role is to sort of govern and administer this whole thing. I may be able to handle even 8 or 9 at a time. But on the other hand I am a PI, and I myself am doing the recruiting and the consent, maybe I can't do more than 2 or 3 at a time. So I don't think there is a numeric cap possible, but the concept of capping is a good one. I think we should say that at any given time you should not be doing more trials than you can. (…) I think it's a good idea to consider that a person can only handle so many at a time.* (Respondent 5, physician and Institutional Ethics Review Board member, June 2012)

He suggests that capping could also be done by limiting the number of trial applications the IERB board can review on a monthly basis; this would ensure that the number of trials does not jeopardize the monitoring process of trial conduct:

> *We meet once a month. What's happening now is that the number of projects that we review every month is rising, so we end up having two meetings a month, then three meetings. For 10-12 members who are full-time professionals in the field to sit*

*for ethical review, I think there should be a maximum number of projects that we can review per month. I think it is right to put a cap on how many they can review at a time.* (Respondent 5, physician and Institutional Ethics Review Board member, June 2012)

To conclude, respondents feel that the Institutional Ethics Review Board can do more to ensure that the process of enrolling patients in clinical trials is done in an ethically sound manner. Capping the number of trials under each PI and capping the amount of trial applications that the IERB reviews every month were suggestions given to increase monitoring capacity. Other solutions voiced are already part of the IERB's requirements as based on its mission statement. For example, checking how many trials a PI is conducting, searching for conflicts of interest, and monitoring the trials are all part of the IERB's stated conditions. From a policy perspective, capping the number of trials could be considered as picking the low-hanging fruit; however, it would be a good start.

### Training and Education in Medical Ethics

As mentioned earlier, much of the responsibility is placed on the PIs to ensure patient understanding, which leads to easy enrollment. Respondents also felt that paying doctors for recruitment is ethically controversial. If physicians can potentially have undue influence on patients entering drug trials, and if pharmaceutical companies offer inducements on recruitment numbers, then the education, training, and awareness of the importance of informed consent becomes paramount. Respondent 5 relates the ethical sense of the PIs and links that to undue inducement:

*And unless they [the PIs] know and more importantly they feel that informed consent is an important thing, it's not going to happen. You may put in any safeguard, but they must feel it also. And that's where the pharmaceutical company dangling inducements to them to recruit patients is a problem.* (Respondent 5, physician and Institutional Ethics Review Board member, June 2012)

Therefore, educating and training doctors in the importance of informed consent was a crucial tool that respondents identified to assure that patients voluntarily give their consent and are fully informed about the details of the trial. Respondent 3 highlights:

*We need to step up case-based education and exposure to what is right and what is not. Doctors here don't know they are doing anything wrong. A large majority will do things and say I don't know it is wrong.* Respondent 3, a professor, physician, and faculty member of medical ethics (July 2012)

High-quality informed consent needs to be placed as a top priority during research; it was felt that this can only be achieved internally through education, awareness, and training. By targeting the education system, respondents felt that an internal moral compass could be triggered. The following quote by respondent 3 portrays this:

> *Here it's amazing you ask them [medical students] questions in the first year, the answers are fantastic, they care for people. Five years down the line you ask them the same questions the answers are different. The medical educator, the clinician, is the one who has to start showing some respect [to patients and their rights]. I don't think it will be an external agency. It has to be internal. That's much harder. What we see now is a reflection of our medical education (…) There is nothing wrong with the doctor, but it is ignorance of my patient's welfare. Why are we here? It's a major educational issue, it's the educational system.* (Respondent 3, professor, physician, and faculty member of medical ethics, June 2012)

Yet, a faculty member of the Department of Medical Ethics, respondent 9, emphasizes how difficult it is to build an internal integrity and individual attitude through training and education, especially when the students' seniors are not providing a good example and are not applying what they have learned:

> *I have been a teacher for 31 years. I have been telling the students to take classes. But one senior is there, your boss … If he doesn't bother to do anything and makes them [patients] sign, if the student is watching that, those five years of teaching won't help. Those five minutes will nullify that. That individual must also do that. At the level of internship medical ethics program we did informed consent. They told us during that, that "so many of our doctors don't bother to do anything." So your four years of teaching will be nullified by those 5 minutes of contact with that doctor. Teach by example.* (Respondent 9, July 2012)

Training on medical ethics and PIs' moral obligations toward their patients can also happen later in the form of good clinical practice (GCP) training. Respondent 6, a professor of medicine and involved in the Department of Medical Ethics, highlights that only some pharmaceutical companies and CROs insist on such training as a way to improve the process of informed consent:

> *One of the ways is to educate the PIs about their responsibilities. Good clinical practice training, they don't all do it. Some multinational companies insist that*

*they be GCP trained. Many of them don't insist.* (Respondent 6, professor of medicine, June 2012)

Others question whether the training reaches their aim, as physicians do not always participate wholeheartedly. Respondent 1 says:

*Yes they run GCP ICH workshops but are they done to satisfy the regulatory departments because they are supposed to do that? How participatory are the doctors and clinicians?* (Respondent 1, professor and member of Institutional Ethical Review Board, June 2012)

Training and education in medical ethics take paramount importance when dealing with local populations that may be vulnerable to inducement by primary physicians acting as principal investigators. The interviewees expressed that there may be conflicts of interest if CROs hire PIs for pharmaceutical companies and offer inducements for high recruitment numbers by physicians.

### Evaluate Patient Comprehension

Most respondents admitted that they do not evaluate the effectiveness of the universalized informed consent procedure. The principal investigators or physicians rarely assess patient understanding before allowing patients to sign the informed consent form. Instead, they commonly give the patient the information sheet to take home and read. Although some details are explained to the patient, the onus is put on patients to educate themselves about the details of the trial before making an informed decision. In light of this, the respondents suggested that the role of the Institutional Ethics Review Board could be expanded from only assessing the documentation of the informed consent to evaluating patient understanding; this would also allow the closing of a research gap. Respondent 6, professor of medicine, points out that there are very few global studies that have evaluated how well a person understands the patient information sheet:

*As far as I know there has not been a study that evaluates how much a person understands. There have been very few studies all over the world that have gone back and seen how much the patient understands about the consent that he has given.* (Respondent 6, professor of medicine, June 2012)

The respondents' level of understanding of the patient information sheet is not often evaluated, especially if the patient signs on the same day. Respondents expressed that PIs assume that patients educate themselves regarding the research details by reading the information sheet before signing the informed consent form. This is in contrast to the fact that sponsors and national regulations expect PIs to explain to patients what is written on the information sheet and ensure full comprehension by the respondents. In fact, the PI may delegate the job of obtaining informed consent to a research coordinator or a nurse. This is permissible under the conduct of clinical trials, but it does not absolve the PI of ultimate responsibility in the ethical conduct of informed consent.

One of the recommendations found in literature concerning informed consent was to test patients on their understanding of the clinical trial procedure prior to signing the consent form in order to prevent misconceptions and to ensure true informed consent. Interestingly, one of the respondents argues that doing this might deter the patient from signing; the patients would become skeptical as to the integrity of the doctor or trial in question if he or she keeps asking whether the information is understood. This attitude was confirmed by Kumar (2008), who states, "Because the patient often considers the physician in such high regard, explaining the different options available, or the possible negative effects of a treatment could make it appear as though the physician is not knowledgeable" (Kumar 2008, 3). In the context of the cultural patient-doctor relationship in India, it seems that evaluation of understanding and testing patients' knowledge of the information provided to them becomes a challenge, which may increase the refusal rate. However, respondent 16, a member of a CRO in Bangalore, says that his company has developed a very short questionnaire of approximately five questions to test patients' knowledge before signing the informed consent sheet:

*We have simplest questionnaire which does not take more than five minutes for them to fill it. Just tick whatever answers they think is right, it takes less than five minutes. So that's another way of monitoring and making sure the patient has understood what we are talking about.* (Respondent 16, member of Contract Research Organization, July 2012)

As can be seen, the respondents expressed concern that before signing a consent form, patients do not fully comprehend the nature or important details of the trial. Unfortunately, this results in a poor informed consent procedure. Therefore, ensuring comprehension by evaluating understanding is one of the solutions that the respondents often articulated.

## Discussion and Conclusion

Referring back to the quote of Christian Byk (2009), the international code of bioethics has conflicting meanings in different local contexts. The aim of this study was to explore the perspectives of physicians and research staff involved in clinical trial conduct on the process of informed consent and the extent to which international bioethical regulations has resulted in effective, voluntary, and fully informed consent. The results demonstrate how actors who are supposed to apply international codes of bioethics struggle when universalized standard procedures are used at a local level. They fear that individual factors, such as the traditional trust between doctors and patients, health literacy, education, language, and socioeconomic status, impede the quality of informed consent. Although long, complicated informed consent forms and patient information sheets represent a barrier themselves, this is further exacerbated by the nature of the patients in question. The results uphold Byk's point in that the diversity of cultures does indeed give a different meaning to ethical principles.

One of the anticipated outcomes was that respondents use innovative ways to deal with their patients and explain the concept of research and clinical trials as a way to offset the difficulties faced when adhering to international standards. Clearly, despite the significant debate regarding the informed consent procedure in relation to the ethics of clinical trials and patient rights and protection, physicians adapt research ethics guidelines to suit local situations when they feel that it is beneficial for the patient. For example, involving family members ensures that patients internalize the details of the trial before they sign the informed consent form. Although international guidelines insist on listing each of the possible side effects of a study, the results of this study suggest that long patient information sheets hamper patient understanding. When doctors explain the trial to their patients, they greatly shorten the information in order to capture the key elements and present those to the patient in a comprehensive manner and in their local language. Physicians dedicate their time to explaining the basic elements of the trial; although international standards dictate that informed consent forms should contain all of the relevant information, local physicians perceive them as a great barrier to getting patients to understand the essentials of the trial. This is another way in which the international guidelines are adapted and contextualized to the local setting. Furthermore, by involving the help of members of the local community, such as local nurses, the language barrier can be overcome, which allows local community members to act as witnesses. Respondents feel that involving the local population in the design of a patient information sheet could increase the quality of the informed consent process by reducing the number of pages, making it more concise, and adjusting it to local needs.

The findings also suggest that participants have several key moral concerns associated with enrolling noneducated patients or those from a low socioeconomic status into a trial. They feared that these patients are more vulnerable to enticement, both as a result of faith in their treating physician and because socioeconomic status has a direct link with what patients perceive as a benefit. For patients in the outpatient department, enrolling in a clinical trial means having quicker access to more thorough clinical care—normally out of their financial grasp—as well as more attention from doctors and free medication. While some doctors only reluctantly admit that patients benefit from free treatment, others clearly argue that the nature of the patients in question undermines voluntariness of trial participation.

Pertaining to the barriers of true consent, one of the leading determining factors is the integrity of the PI. The respondents feel that even when stringent external regulations are applied, it is ultimately the PI's responsibility to ensure effective informed consent. Consequently, respondents suggest that raising awareness, teaching by example during medical school, and good clinical practice training are some recommendations that could be used to enhance PIs' integrity. Furthermore, it is crucial that PIs are monitored by the Institutional Ethics Review Boards, have a cap to the amount of trials they undertake, and are rewarded according to the quality of their work and not the quantity of participants recruited.

The results of this study show that, to a certain extent, international, universal, and standard ethical procedures are being contextualized by how professionals perceive the local needs of patients. However, at the moment, this is being done on a case-by-case basis. Although innovative ways to achieve informed consent using the global standardized process are being used, it is felt that a different approach is needed if regulatory authorities prefer a standardized informed consent procedure. Due to the barriers to obtaining true informed consent, universal bioethics must be indigenized in order to protect patients. This requires engaging the local community in order to design a need-based informed consent process that protects local patients. Such a bottom-up, rights-based, and need-based approach to informed consent will provide the anchor for a reconfiguration of universal bioethics.

These findings have several implications for global health policymakers and international bioethics committees. During the design phase of informed consent forms, the audience (local PIs and trial participants) should be kept in mind. The patients who read the informed consent forms are often daily wage and uneducated workers who may not necessarily be able to understand medical terminology and may not have the time to read lengthy legal documents. Doctors adapt informed consent guidelines on a daily basis to ensure that patients who enroll in clinical trials have a minimum understanding of what the trial involves. They do so using innovative ways such as including family members or enlisting the help of local community members. This adapted method is particularly relevant when dealing with patients from a low socioeconomic status. Patient information sheets should be designed in a local language and only provide the most relevant information. They should move away from being purely legal documents that protect CROs and pharmaceutical companies and become more need-based forms that educate trial participants about their rights. Currently, these documents hardly contribute to increasing patient welfare or the protection of patient rights. Instead, they illustrate a great mismatch between the ideas of policymakers or ethicists and ground-level reality. The results imply that patient rights protection should entail indigenizing bioethics. Policymakers need to look at patient rights in clinical trials from the perspective of trial participants and their needs. The field experience of the respondents suggests that inconsistencies in the governance and enforcement of informed consent procedures might be mitigated if informed consent procedures were indigenized or adapted to local settings. At the same time, the accountability of PIs would be increased by reducing the barriers to applying standardized consent procedures. It would go a long way in overcoming the accountability gap of multinational corporations. If informed consent procedures were adapted to the local culture, language, and sociopolitical environment in which the trial is taking place, the procedure could pave the way for regulatory enforcement.

## Acknowledgments

The authors are grateful to the people who participated in the interviews and provided their valuable time and insights.

## References

Annas, G. J. 2009. "Globalized Clinical Trials and Informed Consent." *New England Journal of Medicine* 360(20): 2050–2053.

Byk, C. 2009. "Chapter 9: The UniversalDeclaration on Bioethics and Human Rights: Bioethics, a Civilizing Utopia in the Age of Globalization? Bioethics: Between Universalism and Globalization." Accessed March 3, 2012 http://www.ashgate.com/pdf/SamplePages/Nexus_of_Law_and_Biology_Pref.pdf.

Department of Health. 2005. *Drugs and Cosmetics (2nd Amendment) Rules.* New Delhi: Ministry of Health and Family Welfare.

Fitzgerald D., C. Marotte, R. Verdier, W. Johnson, and J. Pape. 2002. "Comprehension during Informed Consent in a Less-Developed Country." *The Lancet* 360: 1301–1302.

Kumar, R. 2008. "Evolving Clinical Trials, Pharmaceutical Executive." Accessed March 20, 2014. http://www.pharmexec.com/pharmexec/R%26D/Evolving-Clinical-Trials/ArticleStandard/Article/detail/557676.

LaFraniere, S., M. Flaherty, and J. Stephens. 2000. "The Dilemma: Submit or Suffer, Article 3 of The Body Hunters." *Washington Post.*

Laughton, A. H. 2007. "Somewhere to Run, Somewhere to Hide?: International Regulations of Human Subject Experimentation." *Duke Journal of Comparative and International Law* 18(181): 181–212.

Lee, B. 2010. "Informed Consent: Enforcing Pharmaceutical Companies' Obligations Abroad." *Health and Human Rights* 12(1): 15–27.

Maiti, R., M., R. 2007 "Clinical trials in India", *Pharmacological Research* 56(1) 1–10.

Miller, F. G, and Shorr, A. F. 2002 "Ethical assessment of industry-sponsored clinical trials: a case analysis." *Chest,* 121(4):1337–1342.

Rachid, A. 2006. "Global Clinical Trials in Bangladesh: A Call for Action." *BAPA Journal.* http://www.bapainfo.org/html/documents/1526.pdf.

Rajaraman et al. 2011. "How Participatory Is Parental Consent in Low Literacy Rural Settings in Low Income Countries? Lessons Learned from a Community Based Study of Infants in South India." *BMC Medical Ethics* 12(3). Accessed March 20, 2014. http://www.biomedcentral.com/1472-6939/12/3.

RNCOS. 2007. "Booming Clinical Trials Market in India." Accessed March 20, 2014. http://www.canbiotech.com/CommonData/NewsFiles/Booming%20Clinical%20Trials%20Market%20in%20India.pdf.

Sengupta, S., B. Lo, R. Strauss, J. Eron, and A. L. Gifford. 2011. "Pilot Study Demonstrating Effectiveness of Targeted Education to Improve Informed Consent Understanding in AIDS Clinical Trials." *AIDS Care* 23(11): 1382–1391.

Shah, S. 2006. *The Body Hunters, Testing New Drugs on the World's Poorest Patients.* New York: The New Press.

Silverman, E. 2011. "Clinical Trials Death and Compensation in India." *Pharmalot.* Accessed June 15, 2011. http://www.pharmalot.com/2011/05/clinical-trial-deaths-andcompensation-in-india/.

Strauss et al. 2001. "Community-Based Participatory Research. *American Journal of Public Health* 91(12): 1938–1943.

Strauss, R., P., Sengupta, S., Quinn, S., C., Goeppinger, J., Spaulding, C., Kegeles, S., M., Millett, G. 2001. "The Role of Community Advisory Boards: Involving Communities in the Informed Consent Process, *American Journal of Public Health.* 91(12):1938–1943.

Williams, J. R. 2012. "Exploitation and Developing Countries: The Ethics of Clinical Research." Edited by Jennifer S. Hawkins and Ezekiel J. Emanuel, 327. Princeton and Oxford: Princeton University Press, 2008. The Heythrop Journal, 53: 895–897. doi: 10.1111/j.1468-2265.2012.00757_46.x.

Williams, J. R. (2012), Exploitation and Developing Countries: The Ethics of Clinical Research. Edited by Jennifer S. Hawkins and Ezekiel J. Emanuel . Pp. 327, Princeton and Oxford, Princeton University Press, 2008, $14.95. The Heythrop Journal, 53: 895–897. doi: 10.1111/j.1468-2265.2012.00757_46.x.

Zaman, S., and P. Nahar. 2011. "Searching for a Lost Cow: Ethical Dilemmas in Doing Medical Anthropological Research in Bangladesh." *Medische Antropologie* 23(1). Accessed March 10, 2012. http://tma.socsci.uva.nl/23_1/abstracts.htm.

# CHAPTER 3

# Payments in Clinical Research: Views and Experiences of Participants in South Africa

*Olga Zvonareva and Nora Engel*

## Introduction

Clinical research involving humans introduces multiple ethical controversies, including the one surrounding payments provided to research participants. Providing monetary payments to clinical research participants is a widely used practice, but it continuously raises debates. Concerns have been expressed that payments can influence decisions of potential research participants regarding enrollment in a study, possibly violating voluntariness of consent, "blinding" participants to the risks involved, and leading to undue inducement (McNeill 1997; Lemmens and Elliott 1999; Ballantyne 2008). It has also been hypothesized that by paying participants, investigators may disproportionally attract economically disadvantaged people who come to bear an unfair proportion of the burdens while society at large receives the benefits (McGregor 2005; Phillips 2011; McNeill 1997). Paying participants can also lead to commercialization of participation in clinical research (Chambers 2001; King 2001). According to a long prevailing view, only payments that are intended to reimburse research participants for losses incurred in connection with participation in research, such as travel costs, are considered to be ethically nonproblematic (Wilkinson and Moore 1997).

At the same time, the accepted orthodoxy to keep payments low and avoid sums that could potentially introduce undue inducement and preferential targeting of economically disadvantaged individuals has been increasingly criticized (Anderson and Weijer 2002; Resnik 2008). Critics suggest that such measures are paternalistic, violate autonomy of individuals, and, for the economically disadvantaged, eliminate beneficial and desired possibilities (Wilkinson and Moore 1997). Also, low payments might be unfair and exploitative in relation to the time and effort required for participation in a clinical study (Phillips 2011) while the meaning of "low" is likely to differ for individuals of different socioeconomic standing and is a subject for debate itself (Ballantyne 2008; Dickert and Grady 1999; VanderWalde 2006). Lack of consensus in this regard has resulted in ambiguously formulated policies. There is a clear agreement between various research ethics guidelines that participation in clinical research must be free from undue inducement (Canadian Institutes of Health Research 2010; Council for International Organizations of Medical Sciences (CIOMS) 2002; International Conference on Harmonization 1996; National Health and Medical Research Council 2007; Phillips 2011; The Nuremberg Code 1947; World Medical Association 2008). However, the regulations do not clearly define appropriate amounts and forms of payment, nor do they clarify the line separating fair compensation and undue inducement. They also do not advise how these guidelines should be applied across different settings (Dickert, Emanuel, and Grady 2002). Such ambiguity leads to a great variety of standards with regard to payments in clinical research applied in different practice settings (Dickert, Emanuel, and Grady 2002).

As clinical research expands globally and payments are widely used in recruitment practices, renewed attention is given to this issue (Dickert, Emanuel, and Grady 2002). Clinical trial sites are shifting from established trial regions of Europe and North America to lower-income settings, further increasing concerns as payments could introduce different, potentially more harmful dynamics in various contexts (Petryna 2005, 2009). Against the background of the globalization of clinical research, the controversy of payments accrues additional nuances and uncertainties. Many have worried that the potentially negative effects of payments will be even more pronounced in lower-income settings, where the economically disadvantaged constitute a large part of the population. The rationale behind this concern is that impoverished people, being constrained by their vulnerable circumstances and lack of available options, might see payments offered to clinical research participants as an offer too good to refuse, irrespective of the level of risk involved and degree to which the conditions are fair (Beauchamp et al. 2002).

However, the issue of payments in clinical research has rarely been a topic of empirical inquiry. Notably, perspectives of research participants themselves remain largely invisible in the corresponding ethical debates, taking place mainly among academics and scientists from the economically-rich countries of the West (Stunkel and Grady 2011; VanderWalde and Kurzban 2011). If ethical guidance is to be meaningful on the ground, however, and research conduct is to be in tune with societal expectations, it is essential that research populations engage in discussions on the key ethical question of payments in clinical research. Experiences and motivations of people faced with ethical complexities caused by payments in clinical research can reframe the corresponding debates by exposing overlooked issues and switching focus away from concerns that appear to have little relevance in practice settings. Furthermore, involving trial participants provides insight into how local sociocultural specificities in a particular setting shape the perceptions and attitudes of participants concerning payments in clinical research. This study adds to the growing body of literature calling for research participants' equal inclusion in this debate, as well as in broader discussions about ethics of clinical research for achieving legitimacy and sustainability of global clinical research (Fairhead, Leach, and Small 2006b; Geissler and Pool 2006).

This study focuses on the voices of actual and potential research participants in a low-income setting in South Africa, a transitional country that is attracting a growing numbers of clinical trials. Ethnographic studies suggest that actors (including research participants) in increasingly diverse settings where clinical research is performed actively reinterpret research ethics guidelines and practices to fit local circumstances and value frameworks (Fairhead, Leach, and Small 2006a; Gikonyo et al. 2008; Sariola and Simpson 2011; Stewart and Sewankambo 2010). Thus, to gain an understanding of how payments in clinical research are viewed in South African settings, this study examined perspectives and attitudes of actual and potential research participants on their own terms, as well as their wider backgrounds, ideas, and values underpinning their views.

Audiotaped and transcribed interviews were conducted with 24 individuals, nine of whom had experience participating in clinical research.[1] Initially, only the cleaning staff at one of the South African universities has been invited to participate. This decision was based on the rationale that it is relatively safe to assume the low economic status of these participants, given that their salaries did not exceed the equivalent of $240 per month. Nineteen members of the cleaning personnel were included in the study. As only four of them had actual experience in clinical research, five more individuals were included who were enrolled in a phase III randomized controlled trial designed to assess the safety and effectiveness of a vaginal microbicide in the prevention of HIV Type 1 infections in women. Economic status of the trial participants was not assessed directly. According to the research staff involved in the trial, however, the community in which the trial was conducted has historically been underprivileged and is still rather poor.

Conducting in-depth individual interviews allowed an understanding of why informants expressed certain opinions, the exploration of values and worldviews behind these opinions, and discussions of unforeseen but relevant issues that were brought up by the informants. Over the course of the study, data collection, data analysis, and interpretation were simultaneously performed. During data collection, quality was ensured through review of tapes and transcripts to improve interview methods and content. At the same time, data interpretation and analysis were continuously reviewed and refined. Text units relevant for the study topic were identified and coded using the concepts derived from the data and pre-existing theoretical frameworks. Themes were developed drawing from the codes of similar content and underlying meanings. While analyzing the data, concepts and emerging patterns were cross-checked within and across interviews to challenge or support the informants' and researchers' own accounts and interpretations.

This chapter presents how informants interpreted clinical research, depicts how they perceived payments against the background of their ideas of clinical research, and relates these perceptions to theoretical models of payments to clinical research participants offered in the bioethical literature (reimbursement model, market model, and wage-payment model). The chapter concludes by elaborating on the significance of including voices of research participants in designing bioethical standards for global clinical research conduct.

## "Because We Want to Learn More and Grow"—Views of Informants on Clinical Research

With only few exceptions, informants expressed a positive and welcoming attitude toward clinical research. It was perceived as a search for ways to address health problems prevalent in South African communities. This view is illustrated by the following statement of a woman who, not having the experience of participating in a clinical study, knows several individuals who had been involved in research:

*May be it is in my mind that the research is a very good thing that we can get the clue to know and to get help.* (A, 56-year-old female, no experience in clinical research)

A woman currently enrolled in a microbicide trial echoes this statement, assuming that researchers, just as that woman herself, are motivated to help the South African population:

*Because they are helping us, that is why I want to take part over and over again. Or is a pills, or is a medicine, everything they test, I will go for it.* (S, 31-year-old female, has personal experience in clinical research)

Besides presenting clinical research as locally relevant and beneficial, the informants often stressed the equal importance of clinical research participants and researchers. They commonly referred to everyone involved in the process of clinical research as "we," highlighting its collaborative nature and common goals. For example, explaining why they would take part in a clinical study, the informant said:

> I will agree because we want to search this medication is working ... because we want to grow, to learn more and grow, then at the end we can know this medicine works this way. (Z, 28-year-old female, no experience in clinical research)

This concept of a researcher-participant relationship can be described as a partnership based on the common concern about health in South African communities and a desire to improve it.

The perspectives described in this section are consistent with a growing body of ethics literature advocating social value and collaborative partnerships as important benchmarks of ethical international clinical research (Emanuel et al. 2004; Gikonyo et al. 2008). It is now widely recognized that clinical research should have a capacity for generating social value locally through the production of knowledge that can lead to health improvements (Lairumbi et al. 2012; Molyneux et al. 2012). In the absence of developed infrastructure that translates research results into health system enhancements, as is often characteristic of low- and middle-income localities, individuals and communities assume risks and burdens, while benefits go to researchers, sponsors, and citizens of affluent countries. Actors in academic and policy circles agree that without social value, research that exposes participants to risks might be exploitative (Emanuel et al. 2004; Lairumbi et al. 2012; Molyneux et al. 2012). In this context, special attention is given to collaborative partnerships, which is in line with those mentioned by the participants. These could provide means to minimize exploitation by ensuring that researchers, sponsors, policy-makers, and communities work together to determine health priorities, research designs, and ways to translate research results into practice (Lairumbi et al. 2008).

### "I Can't do It Through Money"—Attitudes Toward Payments in Clinical Research

Some informants thought that payments for participation in clinical research should not be accepted, whereas others took a more neutral position by saying that they would take a financial payment if offered while not actually expecting it. Motivation of gain was commonly portrayed as inappropriate for participation in clinical research.

The view that financial awards were unacceptable was rooted in the perception of local relevance and the value of clinical research. Since the informants conceptualized clinical research as intending to improve health in the places where it is conducted, some of the participants felt it was wrong to accept money for helping their own communities. The story of one informant about her friend's experience illustrates this point of view:

> *They promised they will give her 50 Rands … She refused to take that. Because she understood that this medicine can help a lot of people. If it helps her, it means that most of our nation they will get help. So that is why she refused to take that money. She said: "No you can't pay me, if this medication is not poison, it's a healthy medication, it will be good, because most our family, our friends, everybody in South Africa, they get difference in that.* (A, 56-year-old female, no experience in clinical research)

Other informants did not explicitly reject payments and communicated a somewhat indifferent attitude toward financial reward. They acknowledged that they would accept money when given, but they would not see it as a necessity. This position is reflected in a statement of the youngest informant, responding to the question of what he would expect to be offered as a clinical research participant:

> *Actually I would not say I will expect* [to get anything for participation]. *If they do offer me something, I will accept whatever they give.* (K, 22-year-old male, no experience in clinical research)

Another informant said that if she was invited to participate in a clinical study, she would agree and referred to financial payment specifically:

> *I like money. But even if they don't tell about money I don't have a problem, even there where is no money I can go.* (Z, 28-year-old female, no experience in clinical research)

An informant currently involved in a clinical trial, in which no financial payments were provided except for reimbursement of transportation costs, echoed the perspectives of those who had not participated in clinical studies:

> *For me anything is enough. I am not here for money, to do a research, to help others. For me anything is fine.* (Q, 24-year-old female, has experience in clinical research)

An informant who participated in five small-scale clinical studies conducted by graduate students in an academic hospital explained:

*I do not have a problem whether they give me less money or, maybe more money. Because sometimes even when they give me money, it is not like they are paying me. It is just like may be to let me buy a drink … So even if they don't pay me anything, I do not have a problem.* (C, 38-year-old female, has experience in clinical research)

The common thread running through the interviews was that motivation of gain, including financial gain, was not seen as appropriate for participation in clinical research. One informant compared participation in clinical research with blood donation, claiming that one could accept an award offered but should not be motivated by the prospect of its reception:

*It's like blood donation it's like you feel that you are doing blood donation, it is from your heart. And if you feeling to do it, if government can say that now all the people who are doing blood donation we are going to give them money. So much, what what. Even that it is not a problem. But if you feel to donate you can go.* (G, 32-year-old male, no experience in clinical research)

Another informant compared the desire to gain financially through clinical research to criminal activity:

*Some people would take risk for money. Just like some people who are robbing, they know that it is dangerous, but because that might, it's not 100% sure, gain, they take that kind of risk.* (I, 38-year-old male, no experience in clinical research)

Another informant thought that participating in clinical research after being motivated with the prospect of financial reward could negatively affect the reliability of the results. He was convinced that the appropriate motivation for participating in clinical research was to contribute to the common good:

*I can't do it through money. It won't work because I want something. I can't carry on because of they want to give me something. Yes, I must do it with my heart … If I will do because I want to get something, it is not nice. Tomorrow people end up sick because some things did attract me to do this thing.* (X, 53-year-old male, no experience in clinical research)

Overall, informants with and without experience participating in clinical research viewed the research as an activity beneficial for the health of local populations; they felt that personal rewards for participation, including financial ones, should not be the main motive for participation. These perspectives can shed some light on the findings of a quantitative study conducted in Uganda that explored perspectives of clinical research participants and community members on clinical research benefits and payments (Grady et al. 2008). The authors reported being intrigued by the fact that almost 30 percent of their sample of 915 individuals stated that no personal reward (financial or other kind, such as provision of ancillary care or health items) was necessary in clinical research; the authors recommended further investigation to clarify reasons for this. It is possible that participants in the Ugandan study also viewed clinical research as leading to better health in local communities in the long run and, therefore, did not expect any additional gains. Such perceptions of clinical research were evident in another study conducted in Ghana, which explored the meaning of participation in clinical research among individuals enrolled in malaria vaccine trial (chapter 1).

The views of the informants on clinical research and payments may also reflect local accounts of right and wrong. African moral traditions[2] are relational in nature (Metz 2010; Tangwa 2000; Tangwa 1996). The cornerstone of these traditions is Ubuntu, a traditional philosophy and way of life for communities in South Africa and Sub-Saharan Africa in general (Munyaka and Motlhabi 2009). It is commonly accepted that Ubuntu is derived from the word *muntu*, which means a person, a human being. Ubuntu constitutes the very essence of being human. It is an inner state that makes one feel and behave according to human nature; this involves contributing to the well-being of others and the community and promoting harmonious relations within society (Shutte 2001). Viewing clinical research as a common quest toward better health for society and community coincide with Ubuntu notions of proper human conduct. This would also explain why international clinical research is judged as morally right. Research impartial to local needs and not contributing to the betterment of the host communities (beyond the mere purposes of research) would not have a chance of being accepted. Such research may be viewed as lacking human-ity, violating harmonious relations within society, and morally wrong. Further-more, since clinical research was conceptualized by the informants as an effort to improve health in local communities, research participation was viewed as a commendable contribution into improving the well-being of one's fellow com-munity members. Within the described moral framework, being motivated for such action by the desire of financial gain cannot be considered acceptable.

## Decision-Making: Risk-Benefit Calculus and Effects of Payments

The logic of the risk-benefit calculus as exposed in the interviews some-what departs from the logic established in the theoretical debates, which assumes the possibility that payments might constitute undue inducement that "blind" participants to risks. According to most informants, financial payments had no effect on decision making regarding involvement in clinical research, irrespective of whether there is intention to participate or not. Several informants added that payments might even be seen as indicative of increased risks and low social value of research, thus causing potential research participants to shy away.

Informants unanimously acknowledged that clinical research entails risks for participants. An informant without experience in clinical research described how she would feel taking an experimental drug:

> *I would feel scared, I can't be free, because it's the first test. So it is not going to be easy. I can use it but I won't, I will never say I am safe.* (Z, 28-year-old female, no experience in clinical research)

Informants conveyed that the risk of damaging their health in a context of minimally accessible health care outweighed the personal benefit of obtaining even a considerable sum of money. For example, an informant enrolled in a microbicide trial explained that it was very hard for her to receive medical help:

> *You have to wait long ... public hospitals ... you will go and die there.* (O, 37-year-old female, has experience in clinical research)

Informants in this study attached much value to their own health; in the environment in which they live, it is necessary to be healthy to be able to cope with difficulties introduced by harsh conditions. In their deliberations, they looked further ahead in time than is typically assumed in bioethical arguments, thinking about the devastating consequences of damaging health as a result of participating in risky studies on their ability to provide for themselves and their families, rather than one-time gain.

Contrary to the prominent bioethical concern about poor people feeling compelled to accept an offer to participate in clinical research when being offered cash, the perspectives of informants in this study indicated that the vulnerable circumstances of low-income individuals might actually make them even more aware of the risks in biomedical experimentation. In such circumstances, obtaining money for risking one's health might not be seen as beneficial.

However, while money was not presented as a valid reason to take risks, benefits for one's community could be such a reason. An informant who had not been involved in clinical research previously said, in response to a question at the end of the interview about whether they would participate in a clinical study and why:

> *I am willing to participate in such things. As long as it has a good impact on the other people.* (I, 38-year-old male, no experience in clinical research)

An informant who participated in five clinical studies in an academic hospital stated she would continue participating, while not getting any material rewards:

> *I am still going to continue helping people who need to do researches, because the one who took my blood, she told me that my blood really helps.* (C, 38-year-old female, has experience in clinical research)

An informant enrolled in a microbicide trial explained why she joined the trial:

> *I wanted to take part in something like this. AIDS killing a lot of people and we are looking to stop it.* (R, 32-year-old female, has experience in clinical research)

The local social value of research that benefits the host community becomes introduced in the risk-benefit calculus, playing, at least for some, the central role in the decision-making process and thus being a major motive for participating in clinical studies. There were, however, two informants who expressed a negative attitude toward clinical research, shaped by their previous experiences as research participants. At that time, they felt they were not provided enough information about the studies in which they were enrolled and the results of those studies. These experiences led to suspicions regarding the goals of researchers, and clinical research in general, and a refusal to participate in clinical research in the future. When asked if they would consider participating in clinical research when being offered a considerable sum of money, the two informants did not hesitate with the response:

> *If ever they are testing this medicine I will say no, even if they offer those thousands of Rands or whatever they are offering.* (D, 62-year-old male, has experience in clinical research)

> *I won't do it even if I am poor, I don't think it is a good thing.* (F, 28-year-old male, has experience in clinical research)

Among those with a view of clinical research as a locally beneficial activity, some voiced suspicion about payments. The informants reasoned that if clinical research was supposed to improve health in local communities, then no money should be offered; it is already intended to benefit local people. Therefore, if payments are offered, research might not be intended for the benefit of the populations where it is conducted, and there is no reason to participate in it.

> *No, that's big payments, it's better to go back ... In this bad thing we pay such amount, in these good things we are not paying.* (A, 56-year-old female, no experience in clinical research)

In some informants' views, payments may not only indicate the absence of the described beneficial nature of a research endeavor; it might also reveal that investigators, while pursuing their own goals, take less care of participants or even subject them to danger:

> *How you can pay me money if I am doing thing like this? How can I take part in something and you think to pay me, why you pay me? For what? Or you pay me so I can die and my children live with that money.* (Y, 45-years-old female, no experience in clinical research)

Such perceptions correspond with the findings of another study that showed a positive association between payments in clinical research and perceived risk (Cryder et al. 2010). There are, however, important nuances. For the informants in the present study, payments for clinical research indicated not only increased risk but also low social value for local communities.

Overall, the perspectives put forward by the informants in this study indicate that in some contexts, the perceived prospect of improving health in local communities might provide a stronger motivational drive for participation in clinical research than cash payments. As the two informants who were disappointed in the field of clinical research show, when people suspect clinical studies are not intended to benefit host communities, offers of money are not attractive. Instead, payments might actually work against participation and be interpreted as a warning sign. Finally, the informants viewed research and its results as public goods. The ethical concern about commercializing participation in clinical research, which has arisen in part due to the appearance of professional "guinea pigs"—a phenomenon described in Western countries (Lemmens and Elliott 2001)—might not hold much relevance in these local communities.

## Discussion and Conclusion

The perspectives of the informants in this study represent implications for some of the models of payment developed in the field of bioethics: the reimbursement model, market model and wage-payment model (Dickert and Grady 1999; Resnik 2008). The reimbursement model stipulates that payment is provided simply to cover participants' expenses. It holds that research participation is not considered paid labor or a tradeable good; it is a type of public service. The informants in this study currently enrolled in the microbicide clinical trial acknowledged the convenience of getting reimbursed for transportation to the research center:

*People who participate they don't stay in the same location. So it helps because they give you a day and tell you must come on a certain date. And if you don't have money, it would be another problem. So they think of us.* (P, 28-year-old female, has experience in clinical research)

However, aspirations of the informants exceeded the frame of this model in that participation was not presented as a gift with no strings attached. Expectations of informants were not limited to getting reimbursed; they would rather contribute to outcomes of clinical studies to be used for the benefit of host communities.

*Some things, especially like the research, is the best. Because at the end we all gain and we grow up. This is our lives.* (Z, 28-year-old female, no personal experience in clinical research)

The market model allows supply and demand to determine whether and how much research participants are paid. According to this model, participants are paid for services (e.g., undergoing procedures) and goods (e.g., providing biological samples for investigations). Large payments could be used as incentives to enroll people in riskier and/or more arduous studies with little prospects of direct medical benefit.

The wage-payment model proposes to standardize amounts paid for participation based on the level of hourly payment for unskilled labor, thus presenting participation as a job. Motivations of the informants in this study and their perceptions of research participants' roles did not conform to the last two models; the informants expressed their desire to contribute to the common good and not to enroll in clinical research for obtaining financial rewards. The feeling of personal responsibility to contribute to the presumably socially relevant activity of clinical research was conveyed by many informants. For example, an informant who has never been invited to participate in clinical research described how she would make a decision if she found a clinical study for which she was eligible:

*I must do it just to help people, even if I don't receive anything.* (L, 42-year-old female, no personal experience in clinical research)

Informants also viewed research participants as contributors and partners in the research process and not just sources of information, samples, or hired workers. Overall, relational understanding of clinical research and payments within the trials exhibited by the informants diverted from the rhetoric of individualism implicit in influential models of payment offered in bioethics.

Against the background delineated in the previous sections, this study demonstrates that various concerns over ethically problematic effects of payments—including the possibility of undermining voluntariness of consent, blinding participants to risks, and commercializing participation in clinical studies—might not be universally relevant in all socially and culturally diverse settings in which clinical research is presently being conducted. Informants generally explained that in viewing research as a public good, they would not expect to be paid; to a large extent, their motivation would depend on the perceived prospects of health benefits for their communities. Furthermore, they explained that for a trial in which they had no desire to participate (when doubts arise about locally beneficial nature of a clinical study), no amount of money would lure them into participating in experimentation. On the contrary, offers of money have the potential to negatively impact levels of participation, if these offers are interpreted as signs of danger. Informants themselves, when speaking about clinical research, emphasized its social value and the promises it holds for local health. They did not focus on payments, nor did they even introduce the topic of payments until specifically asked. Consequently, the dimension of fairness, which often arises in discussions about payments in clinical research, accumulates a broader meaning in light of this study's results. It becomes intimately connected with the more general concept of clinical research benefits (Zvonareva et al. 2013); all perspectives on financial payments articulated in this study were grounded in the hope that clinical research would bring better health to local communities.

Benefit sharing, as a key dimension of ethical research conduct and a means to minimize exploitation, has been articulated in academic and policy circles (Ballantyne 2010; Participants in the 2001 Conference on Ethical Aspect of Research in Developing Countries 2004). Although disagreement persists about how benefit sharing should be organized in practice (Schroeder and Gefenas 2012), general agreement exists about the need to engage host communities while deciding what kinds of research benefits should be provided and to whom these benefits should be given (Emanuel et al. 2004; Lairumbi et al. 2012). This study's results highlight this benefit-sharing framework and stress the necessity of dialogue with local communities for achieving fair benefit distributions. The study informants expected partnerships with researchers and use of research results for the benefit of host populations. It also shows that the ethical challenges of payments in clinical research cannot be viewed and discussed separately from wider discussions about research benefits.

Finally, this study demonstrates how local moral accounts of right and wrong shape participants' interpretations and attitudes toward clinical research, including the issue of payments to research participants. Dominant ideas in research ethics expressed in standardized guidelines, harboring notions of individualism and autonomy, largely draw on Western moral philosophy. Different ways in which clinical research is framed and interpreted within particular cultural horizons and actor groups are not fully recognized in these bioethical arguments. For the normative guidance with regards to payments to be adequate and meaningful, it is important to take into account local values and systems of meaning. The development of ethical guidelines and standards in this area cannot be separated from social realities and circumstances where clinical studies are being conducted. Local sociocultural specificities shape the way clinical research and related payments are viewed. This needs to be recognized by those involved in creating bioethical standards for global clinical research conduct in order to avoid antagonizing societal expectations and establish and sustain community relations that are necessary for continuous operation of clinical research globally.

## Notes

1. Results of this study pertaining to clinical research benefits were published in O. Zvonareva,, N. Engel, A. Dhai, R. Berghmans, E. Ross, and A. Krumeich, 2013, "Engaging Diverse Social and Cultural Worlds: Perspectives on Benefits in International Clinical Research from South African Communities," *Developing World Bioethics*, doi:10.1111/dewb.12030.

2. Scholars, drawing on anthropological inputs and their own cultural backgrounds as members of indigenous communities have described common uniting traits in the rich and long-standing ethical heritage of Africa with its 50 countries and more than a thousand languages. See, for example, works by Metz (2010) and Tangwa (1996).

## References

Anderson, J. A., and C. Weijer. 2002. "The Research Subject as Wage Earner." *Theoretical Medicine and Bioethics* 23(4–5): 359–376.

Ballantyne, A. 2008. "Benefits to Research Subjects in International Trials: Do They Reduce Exploitation or Increase Undue Inducement?" *Developing World Bioethics* 8(3): 178–191. doi:10.1111/j.1471-8847.2006.00175.x.

Ballantyne, A. J. 2010. "How to Do Research Fairly in an Unjust World." *The American Journal of Bioethics: AJOB* 10(6): 26–35. doi:10.1080/15265161.2010.482629.

Beauchamp, T. L., B. Jennings, E. D. Kinney, and R. J. Levine. 2002. "Pharmaceutical Research Involving the Homeless." *Journal of Medicine and Philosophy* 27(5): 547–564.

Canadian Institutes of Health Research. 2010. "Tri-Council Policy Statement: Ethical Conduct for Research Involving Humans." http://www.pre.ethics.gc.ca/pdf/eng/tcps2/TCPS_2_FINAL_Web.pdf. Accessed on June 21, 2014.

Chambers, T. 2001. "Participation as Commodity, Participation as Gift." *The American Journal of Bioethics: AJOB* (1:2): 48.

Council for International Organizations of Medical Sciences (CIOMS). 2002. *International Ethical Guidelines for Biomedical Research Involving Human Subjects.* Geneva, Switzerland: World Health Organization.

Cryder C. E., A. J. London, K. G. Volpp, and G. Loewenstein. 2010. "Informative Inducement: Study Payment as a Signal of Risk." *Social Science and Medicine* 70(3): 455–464.

Dickert, N., and C. Grady. 1999. "What's the Price of Research Subject? Approaches to Payment for Research Participation." *The New England Journal of Medicine* 10(3): 198–203.

Dickert, N., E. Emanuel, and C. Grady. 2002. "Paying Research Subjects: An Analysis of Current Policies." *Annals of Internal Medicine* 136(5): 368–373.

Emanuel, E. J., D. Wendler, J. Killen, and C. Grady. 2004. "What Makes Clinical Research in Developing Countries Ethical? The Benchmarks of Ethical Research." *The Journal of infectious diseases* 189(5): 930–937. doi:10.1086/381709.

Fairhead, J., M. Leach, and M. Small. 2006a. "Where Techno-Science Meets Poverty: Medical Research and the Economy of Blood in The Gambia, West Africa." *Social Science and Medicine (1982)* 63(4): 1109–20. doi:10.1016/j.socscimed.2006.02.018.

Fairhead, J., M. Leach, and M. Small. 2006b. "Public Engagement with Science? Local Understandings of a Vaccine Trial in the Gambia." *Journal of Biosocial Science* 38(1): 103–16. doi:10.1017/S0021932005000945.

Geissler, P. W., and R. Pool. 2006. "Editorial: Popular Concerns about Medical Research Projects in Sub-Saharan Africa—a Critical Voice in Debates about Medical Research Ethics." *Tropical Medicine and International Health* 11(7): 975–982. doi:10.1111/j.1365-3156.2006.01682.x.

Gikonyo, C., P. Bejon, V. Marsh, and S. Molyneux. 2008. "Taking Social Relationships Seriously: Lessons Learned from the Informed Consent Practices of a Vaccine Trial on the Kenyan Coast." *Social Science and Medicine (1982)* 67(5): 708–720. doi:10.1016/j.socscimed.2008.02.003.

Grady, C., J. Wagman, M. J. Wawer, D. Serwadda, M. Kiddugavu, F. Nalugoda, R. H. Gray, et al. 2008. "Research Benefits for Hypothetical Trial: The Views of Ugandans in the Rakai District." *IRB Ethics and Human Research* 30(2): 1–7.

International Conference on Harmonization. 1996. "Guideline for Good Clinical Practice Trial (Vol. 1996)." http://www.ich.org/fileadmin/Public_Web_Site/ICH_Products/Guidelines/Efficacy/E6_R1/Step4/E6_R1__Guideline.pdf. Accessed on June 21, 2014.

King, N. 2001. "Treating Research Subjects as Unskilled Wage Earners: A Risky Business." *The American Journal of Bioethics* 1(2): 53–54.

Lairumbi, Geoffrey M., M. Parker, R. Fitzpatrick, and M. C. English. 2012. "Forms of Benefit Sharing in Global Health Research Undertaken in Resource Poor Settings: A Qualitative Study of Stakeholders' Views in Kenya." *Philosophy, Ethics, and Humanities in medicine: PEHM* 7: 7. doi:10.1186/1747-5341-7-7.

Lairumbi, Geoffrey Mbaabu, S. Molyneux, R. W. Snow, K. Marsh, N. Peshu, and M. English. 2008. "Promoting the Social Value of Research in Kenya: Examining the Practical Aspects of Collaborative Partnerships Using an Ethical Framework." *Social Science and Medicine (1982)* 67(5): 734–747. doi:10.1016/j. socscimed.2008.02.016.

Lemmens, T., and C. Elliott. 1999. "Guinea Pigs on the Payroll: The Ethics of Paying Research Subjects, Accountability in Research: Policies and Quality Assurance." *Science* 7(1): 3–20.

Lemmens, T., and C. Elliott. 2001. "Justice for the Professional Guinea Pig." *The American Journal of Bioethics* 1(2): 51–3.

McGregor, J. 2005. " 'Undue Inducement' as Coercive Offers." *The American journal of bioethics: AJOB* 5(5): 24–25.

McNeill, P. 1997. "A Response to Wilkinson and Moore Paying People to Participate in Research: Why Not?" *Bioethics* 11(5).

Metz, T. 2010. "African and Western Moral Theories in a Bioethical Context." *Developing World Bioethics* 10(1): 49–58. doi:10.1111/j.1471-8847.2009.00273.x.

Molyneux, S., S. Mulupi, L. Geoffrey, and V. Marsh. 2012. "Benefits and Payments for Research Participants: Experiences and Views from a Research Center on the Kenyan Coast." *BMC Medical Ethics* 13(13). doi:10.1186/1472-6939-13-13.

Munyaka, M., and M. Motlhabi. 2009. "Ubuntu and Its Socio-Moral Significance." In *African Ethics. An Anthropology of Comparative and Applied Ethics*, edited by M. F. Murove. University of KwaZulu-Natal.

National Health and Medical Research Council. 2007. *National Statement on Ethical Conduct in Research Involving Humans.* Commonwealth of Australia. http://www. nhmrc.gov.au/_files_nhmrc/publications/attachments/e35.pdf. Accessed June 21, 2014.

Participants in the 2001 Conference on Ethical Aspect of Research in Developing Countries. 2004. "Moral Standards for Research in Developing Countries from 'Reasonable Availability' to 'Fair Benefits.'" *The Hastings Center Report* 34(3): 17–27.

Petryna, A. 2005. "Ethical Variability: Drug Development and Globalizing Clinical Trials." *American Ethnologist* 32(2): 183–197.

Petryna, A. 2009. *When Experiments Travel: Clinical Trials and the Global Search for Human Subjects.* Princeton: Princeton University Press.

Phillips, T. 2011. "Exploitation in Payments to Research Subjects." *Bioethics* 25(4): 209–219. doi:10.1111/j.1467-8519.2009.01717.x.

Resnik, D. B. 2008. "Increasing the Amount of Payment to Research Subjects." *Journal of Medical Ethics* 34(9): e14. doi:10.1136/jme.2007.022699.

Sariola, S., and B. Simpson. 2011. "Theorizing the 'Human Subject' in Biomedical Research: International Clinical Trials and Bioethics Discourses in Contemporary Sri Lanka." *Social Science and Medicine (1982)* 73(4): 515–521. doi:10.1016/j. socscimed.2010.11.024.

Schroeder, D., and E. Gefenas. 2012. "Realizing Benefit Sharing—the Case of Post-Study Obligations." *Bioethics* 26(6): 305–14. doi:10.1111/j.1467-8519.2010.01857.x.

Shutte, A. 2001. *Ubuntu: An Ethic for the New South Africa.*" Cape Town, South Africa: Cluster Publications.

Stewart, K., and N. Sewankambo. 2010. "Okukkera Ng'omuzungu (Lost in Translation): Understanding the Social Value of Global Health Research for HIV/AIDS Research Participants in Uganda." *Global Public Health* 5(2): 164–180.

Stunkel, L., and C. Grady. 2011. "More than the Money: A Review of the Literature Examining Healthy Volunteer Motivations." *Contemporary Clinical Trials* 32(3): 342–352. doi:10.1016/j.cct.2010.12.003.

Tangwa, G. 2000. "The Traditional African Perception of a Person Implications." *Hastings Center Report* 30(5): 39–43.

Tangwa, G. B. 1996. "Bioethics: An African Perspective. *Bioethics* 10(3): 183–200.

The Nuremberg Code. *Trials of War Criminals before the Nuremberg Military Tribunals under Control Council Law (1949)* 10(2): 181–182. http://www.state.nj.us/health/irb/documents/nuremburg_code.pdf. Accessed June 21, 2014.

VanderWalde, A. 2006. "Undue Inducement: The Only Objection to Payment." *The American Journal of Bioethics: AJOB* 5(5): 25–27. doi:10.1080/152651605002 45048.

VanderWalde, A., and S. Kurzban. 2011. "Paying Human Subjects in Research: Where Are We, How Did We Get Here, and Now What?" *The Journal of Law, Medicine and Ethics : A Journal of the American Society of Law, Medicine and Ethics* 39(3): 543–558. doi:10.1111/j.1748-720X.2011.00621.x.

Wilkinson, M., and A. Moore. 1997. "Inducement in Research." *Bioethics* 11(5): 373–389.

World Medical Association. 2008. "World Medical Association Declaration of Helsinki." *European Journal of Emergency Medicine: Official Journal of the European Society for Emergency Medicine* 8. http://www.wma.net/en/30publications/10policies/b3/17c.pdf. Accessed June 21, 2014.

Zvonareva, O., N. Engel, A. Dhai, R. Berghmans, E. Ross, and A. Krumeich. 2013. "Engaging Diverse Social and Cultural Worlds: Perspectives on Benefits in International Clinical Research from South African Communities." *Developing World Bioethics.* doi:10.1111/dewb.12030.

# CHAPTER 4

# Community Health Workers in a Community-Based Tuberculosis Program: Linking Different Social Worlds

*Phuong Nguyen Thi Mai and Nora Engel*

## Introduction

Tuberculosis (TB) remains one of the most prevalent infectious diseases in the world, resulting in relatively high incidences of both mortality and morbidity. TB is curable, yet treatment requires strictly administered long-term intermittent therapy. Eliminating the disease has been a great challenge, despite many years of international efforts (Gabriel 2011). Some of the reasons for this include increasing populations; poverty; poor availability of health care; increasing multidrug resistance as a consequence of treatment failure or poor compliance; emerging HIV coinfection; and delays in diagnosis (Gabriel 2011; Ayisi et al. 2011). In many countries, National Tuberculosis Programs (NTPs) are increasingly turning to other health service providers, including the communities in which patients live, to improve the delivery of effective TB care. Efforts to encourage community contribution for effective TB control have been made, particularly in Sub-Saharan Africa, where the HIV/AIDS epidemic has seriously exacerbated the TB situation (Maher 2003). In 1996, the World Health Organization (WHO)-coordinated project "Community TB Care in Africa" was introduced in Sub-Saharan Africa; this project evaluated the community contribution to effective TB control, as part of NTP activities, in eight district-based projects in six high-HIV prevalent countries (Botswana, Kenya, Malawi, South Africa, Uganda, and Zambia) (WHO 2000). As a result, countries like Kenya introduced Community-Based TB (CB-TB) programs.

Community Health Workers (CHWs) play a crucial role in the implementation of effective CB-TB programs (Ndege and Gitau 2007). CHWs are usually community members who work voluntarily (Singh et al. 2011), although they are sometimes given support and facilitation such as training allowances or transportation reimbursement (Ndege and Gitau 2007). CHWs are not only the first point of contact for the community with the health system, but are also considered to be connectors between health care consumers and providers (Bhattacharyya et al. 2001). CHWs are particularly important in underprivileged and marginalized communities where people may have limited resources, limited access to health care, or different cultural beliefs. As such, CHWs play an integral role in helping systems become more relevant to the people they serve. Common duties include identifying TB cases, tracing defaulters, in-home follow-up of cases, and providing health education (Ogumu 2011). Recognition of the roles, skills, and contributions of CHWs to TB control programs, including a thorough understanding of CHWs' situations through a focus on their personal experiences, will support policy makers and program planners in more effectively approaching target populations.

This study aims to gain a deeper understanding of the role of CHWs in the TB program in a local dispensary in Kenya. It focuses on the work and position of CHWs, discusses the challenges with which they are confronted, and reflects on possible ways to overcome these challenges. In doing so, it reveals the work done by CHWs to make standardized disease controls function in a local community.

Kenya ranks 13th out of the 22 countries with the highest TB burden in the world (WHO 2010). In 1980, the National Leprosy and Tuberculosis Program (NLTP) was established; by the end of 2005, TB and leprosy services were being delivered through 1,605 health units managed by the Ministry of Health (MOH) and other stakeholders, which included NGOs, faith-based organizations, and other private organizations (Ndege and Gitau 2007). At each level, the health unit is responsible for the management, coordination, supervision, and technical advice of controlling TB. It sends periodic reports to the NLTP (Ndege and Gitau 2007). In an effort to improve health service delivery, the Second National Health Sector Strategic Plan proposed six levels of care delivery, ranging from community to tertiary care hospitals (MOH 2007). Three essential categories of personnel at the community level outlined in this plan are CHWs, Community Health Extension Workers (CHEWs), and members of the Community Health Committee (CHC). A CHEW is a trained health professional with a certification in nursing or public health who supervises CHWs and works with the CHCs to ensure accountability. A CHC is a group of people who have the responsibility of leading health action at the community level. The CHWs report to the CHC through the CHEW; the CHEW also acts as the secretary of the CHC (MOH 2007). The CB-TB program was piloted in Machakos in 1997 and was found to be a viable and cost-effective intervention for TB control (Niganda, Wang'ombe, and Floyd 2003). Since then, the Ministry of Health, together with the Center for Disease Control and Prevention (CDC/KEMRI), have involved CHWs or Community TB Ambassadors (CTBAs) in CB-TB programs, allotting them the responsibility for TB suspects' identification, case referral, tracing defaulters, and the provision of health education (Ogumu 2011).

The primary participants of the study were CTBAs who were directly attached to the Kogelo Dispensary in the Siaya District of the Nyanza Province, Kenya. Nyanza is a very poor province; more than two-thirds of the population live below the poverty line (Ayisi et al. 2011). High rates of TB-related morbidity and mortality have been reported for the province; the province has the highest number of smear-positive pulmonary TB cases in Kenya (Ndege and Gitau 2007). Deaths among TB patients range from eight to fifteen percent, with seventy-three percent of patients being coinfected with HIV. Siaya District has three hospitals, fifteen health centers and nineteen dispensaries. The doctor to patient ratio is 1:96,000, and the average distance to a health facility is six kilometers (Ogumu 2011). In Siaya, the CB-TB strategy was introduced in 2004 with the goal of empowering community members to improve the management of TB/HIV at a village level (Ogumu 2011). The CHWs were trained in 2009 to work specifically in the CB-TB program. According to the Kogelo Dispensary monthly report of May 2012, the Kogelo Dispensary covers ten villages consisting of 1,173 households and a total population of 6,130. Currently, 13 CHWs are officially attached to this dispensary; others are doing the work of CHWs but are not formally included in the government's scheme. In addition to in-depth interviews with CHWs and other observations, supplementary information was collected through in-depth interviews with health personnel in the Kogelo Dispensary, including clinicians, nurses, CHEWs, CHC members, and TB patients.

In order to gain a deeper understanding regarding the work of CHWs in the TB program, their relationships, and interactions with people and things, Strauss's "social worlds" framework (Clarke and Star 2008) was used. Social worlds are "groups with shared commitments to certain activities, sharing resources of many kinds to achieve their goals and building shared ideologies about how to go about their business" (Clarke and Star 2008). In their daily lives, individuals often participate in different, often closely related, social worlds (e.g., household, employment, or school; as a patient, as a member of a social organization, etc.). Star and Griesemer(1989) use the concept of "boundary objects" to explain how things and people at the junctures of different social worlds can manage with both diversification and cooperation (Clarke and Star 2008). Boundary objects are "objects which are both plastic enough to adapt to local needs and the constraints of the several parties employing them, yet robust enough to maintain a common identity across sites" (Clarke and Star 2008). A CB-TB program is a complex activity involving different actors that belong to different social worlds—such as patients, practitioners, and communities—at different levels (central to grassroots level). Understanding CHWs as connectors or boundary objects that link and mediate the work at the grassroots level, such as between health care providers and patients of the TB program, helps to explain the involvement, position, roles, and contributions of the CHWs in operating such a program.

Following the social worlds and boundary objects framework, the study also uses Clarke's "situational analysis" approach for analysis of the data. Clarke recommends using three different types of maps to interpret local settings in a varied manner to enable "relational analysis": situational maps, social worlds/arenas maps, and positional maps (Clarke and Star 2008).

In situational maps, all actors (individual or collective) and actants/nonhumans (elements, bodies, and discourses) are mapped; their relationships to each other are then analyzed (relational analysis) (Clarke and Star 2008). Maps of social worlds/arenas take hold of different "universes of discourse"; they map collectives and "sites of action" (Clarke and Star 2008). Based on this type of mapping, interactions and power relations between different actors can be viewed from diverse angles (Clarke and Star 2008). The social arena incorporates different social worlds that have relations with each other. In this study, the Kogelo Dispensary can be seen as an example for a social arena; the health facility continues functioning through complex interactions between different social worlds of health personnel, CHWs, and patients. In order to function smoothly, something or someone needs to take the role of mediator or boundary object, acting at the juncture. This boundary object is flexible enough to fit into either side, as well as harmonize the work of others to achieve a common goal. Positional maps are designed to grasp the different positions taken by different actors in the field (Clarke and Star 2008).

These maps help present a situation from the different viewpoints of the actors involved and avoid a limited view on a particular issue. In this study, the positional maps showed the position of the CHWs in the TB program with regard to their community, the TB patients (they might have been patients themselves or may have relatives infected with TB), and the health facility in which they were trained. It became clear that CHWs do not occupy a fixed position; this helps them be flexible and adaptable at the juncture between different social worlds. Finally, combining the maps helps illustrate different social worlds and their relations within each arena, as well as identify the positions and roles of boundary objects. Through data mapping (using these three types of maps), elements that are related to the CHWs' situations could thus be tracked and analyzed in their complex relations. The analysis highlights that CHWs are the key to the success of a community-based health programs; they can effectively link different social worlds together. Yet, the requirements of CHWs to successfully connect the community with standardized disease control programs and health practices differ in place and time. The resulting implications for policy makers and program managers call into question standardized approaches to community-based health programs.

## Characteristics of CHWs/CTBAs in the CB-TB Program

In order to work as a CHW, each person must first be nominated by the chief or assistant chief and CHC members of the village, based on selection criteria from MOH's "Community Strategy Guidelines" (MOH 2007). Following this, the candidate must be accepted by the community during a village meeting. Most CHWs in Kogelo were chosen about 20 years ago when the Kenyan government started CHW programs throughout the country. This section discusses their duties, motivations, and struggle with numerous challenges.

### CHWs/CTBAs' Duties and Motivations

According to government guidelines, a CHW should be the one concerned about the welfare of the people, respected by villagers as an example of healthy behavior, a permanent resident in the area, and willing to work voluntarily (MOH 2007). In addition, most CHWs add that the "ability of keeping a secret" is a crucial characteristic. CHWs must build trust with the community, especially when they work with sensitive diseases such as TB and HIV/AIDS. Moreover, the reputation of each CHW plays an important role in patients' decision of whether or not to trust the CHW.

> You must keep secret … You must be active to your patºients … The other characteristic a CHW needs is cleanliness, you must be clean. (CHW/CTBA 05—Recording 120612_005)

> First of all, a good CHW should be discipline … They should be confidential, keeping secret … Another thing is they should do certain examples in the community, like role models, by examples, they can teach other people. (CHEW— Recording 120615_001)

> This person, she should be someone from that community where she's attending to. Why? You can't pick someone from other places … no, just somebody from that

*village, at that village level with local language, who can offer much to them, can give them what they need … who can communicate with them well. (Clinician— Recording 120613_003)*

CHWs have developed health-related knowledge and skills through participating in trainings sponsored by different organizations. Topics of those trainings range from prevention, identification, and treatment of common diseases—such as malaria, pneumonia, worms, skin infections, HIV/AIDS, and TB—to pregnancy, immunization, and children's health. In addition, they have been trained in concepts such as development, community mobilization and participation, leadership, communication and practical skills of data recording, and writing reports and proposals. CHWs are equipped with basic knowledge about the signs and symptoms of TB and its cure. During weekly health talks at the health facility or home visits, they educate and mobilize the community regarding TB. The duties of a CTBA, a CHW specifically trained for TB, include referring TB suspects to a health facility, tracing defaulters in villages, and providing health education. At times, CTBAs have to act as TB patients' caretakers. When patients are weak, bedridden, and left without a caregiver, the CTBAs change their bedding and clothes, prepare their food, and make sure that the patients take their TB drugs in the correct manner.

*If there's something needed to be prepared, like tea, like porridge, you prepare and give the patient, then he can take drugs, it is very tiresome but we do it if the patient has no caretaker. (CHW/CTBA 02—Recording 120611_001)*

The CHWs/CTBAs mentioned that the first reason for them to become a CHW is to the benefit their community. They are very proud of being chosen by their village and perceive their contributions to the community to be of great importance.

*She was impressed before with the work of CHW who were doing, so she decided to become a CHW to commit herself to her community so that the community can get better things. (CHW/CTBA 10—Recording 120614_005—Translator 01)*

Recognition of the CHWs' contributions by the government and the community encourages them to continue their work as volunteers. Seeing changes in the community over time, not only in health indicators but also in ways of thinking and living, makes the CHWs feel proud and motivated.

*In the community, you are the one they look for, even you're tired they still call you because the community, they know you, you are their counselor in the village. (CHW/CTBA 05—Recording 120612_005)*

Moreover, the knowledge that CHWs/CTBAs gain from training and the community encourages them to continue working for many years. They also receive valuable advice from people in their community, such as from the elderly or health workers within the facility. Support and encouragement from their families also plays an important role in motivating the CHWs to continue this voluntary work, which requires a great deal of their time and effort.

> *That's education for me because I haven't seen a patient with leprosy before. You can get new things when you walk around the village. Other knowledge you can get, you can get from elder people. (CHW/CTBA 05—Recording 120612_005)*

> *My wife, she's been helping me and I believe I'm here because of her, if it's not her, I'll be somewhere else, not here, searching for something to eat or so. (CHW/CTBA 11—Recording 120614_002)*

Most of the CHWs/CTBAs did not directly mention incentives as the motivation for their work. Sometimes they do get transportation reimbursement or bread and milk for lunch after attending training. However, the hope of "getting something," including new opportunities for further professional development, encourages them to continue working.

> *She managed to get a chance as a Village Reporter, from there now she can be given something small to bring home; that's a very good opportunity for her. She had a certificate, she's hoping for the future, maybe she can get an employment, maybe an NGO, she can work and get something at the end of the day. (CHW/CTBA 07—Recording 120613_002—Translator 01)*

### CHWs' Struggle to Overcome their Challenges

CHWs/CTBAs encounter numerous challenges in their work from the community, their personal situations, and the nature of their job. The community, for instance, may resist their efforts or object to their presence. Many TB suspects refuse to go to the health facility to be tested or may drop out of treatment. The reasons for this include the strong side effects of the medication that cannot be taken on an empty stomach and the rumors and stigma associated with TB due to high rates of TB-HIV coinfection. The most important reason, however, is poverty. Patients have no money for transport, or they have no one to take them to the hospital.

> *The patient is sick at home; his family doesn't have even a cent, you go and get a motorbike by yourself, you bring the person here, the person's being taken care of, it's your responsibility again to take the sick person back home, the family at*

*times, they draw back, they don't want to attend. (CHW/CTBA 05'—Recording 120613_001—Translator 01)*

Working with the community means dealing with diverse characteristics, different ways of thinking, and people from various backgrounds. Misunderstandings and doubts might lead to distrust and refusal. For example, some people believe that CHWs are paid for their work and receive something for referring patients. Others might think that CHWs are just walking around the village because they have nothing else to do. Many villagers are not at ease with a CHW from the same village, either because they believe "that person is just like me, she knows nothing" or because of being afraid that their secrets will not be kept:

*They don't believe that I can tell the exact thing, they think I might be one of gossipers. (CHW/CTBA 02'—Recording 120611_003)*

Difficulties come not only from the community but also from CHWs' own personal situations. Most of them have large families with many dependent members; the number of children they have range from four to ten. In addition to their own children, they often have deceased family members' children to raise. Poverty is the most common problem in the area for patients as well as CHWs themselves.

*Since they're not paid, work is very difficult because they leave a lot of their work with their family. We always tell them to be role models; you can only be a role model if you have something in hands, something to put on the table, but it's not there. (CHEW—Recording 120615_001)*

Their workload and working conditions are additional challenges. CHWs have to combine the tasks of CHWs and CTBAs during their field work, not only focusing on TB but also other health–related problems. Each CHW officially covers ninety households consisting of about five hundred people. Door-to-door home visits and confidential conversations in each household are necessary to identify sick people and convince them and their caregivers to go to the health facility. All of the workers conduct their visits on foot; the distances, poor road conditions, and variable weather conditions can challenge them to successfully complete their work.

*They walk everywhere with their bags, maybe a simple plastic bag with only a bottle of water or a bag with logo and slogan of an NGO's program, inside there are a referral form and pens. (Notes—11/06/12)*

Because of the long period that they have been working as CHWs and CTBAs, each of them has their own way of doing the job and tackling the various problems that emerge in the community. After completing their household chores in the morning, CHWs start their work around noon. On average, a CHW spends three to four hours walking around the village and talking to people. The timing of this depends on other duties of the CHWs and the villagers. It cannot be early in the morning, as most people are tending to their gardens; it cannot be late in the afternoon, as people go to the market to sell their products.

Convincing TB suspects to go to the health facility is a difficult task; CTBAs must be well-prepared and very patient. Through repeated visits, talking slowly, and sharing their own personal stories and experiences, they are able to gain the trust of patients and their caregivers. For difficult and sensitive cases, their approach needs to be very flexible; sometimes they have to call other CHWs or the CHEW for help.

> You should be ready for anything in the field because all those things are in the community. (CHW/CTBA 07—Recording 120613_002—Translator 01)

> By the time you're told to be away by the patients, don't be away ... just be patient! (CHW/CTBA 03—Recording 120611_002)

> She can tell me her story; and me, I also tell her mine and we can go together. (CHW/CTBA 06—Recording 120612_006)

> The community we're working in, people are not the same, some are rich, some are poor, if you go to a poor man's house, don't show them that you're rich and the person's poor, fit in the situation you find there! (CHW/CTBA 07—Recording 120613_002—Translator 01)

In summary, CHWs/CTBAs in the CB-TB program face numerous challenges from the work itself, the community, or their own personal situations. By adapting to the timing of people in the community and flexibly adjusting to each patients' situation, they try to identify and refer TB suspects, trace TB defaulters, and provide health education to the community.

## Linking Different Social Worlds

Health workers, TB patients and their caregivers, CHWs/CTBAs, TB program planners, organizations, institutions, and practices involved in the CB-TB program constitute a social world, within which they interact to achieve one common goal: reducing the burden of TB in the community. At the same time, away from the social world of the CB-TB program, each of these groups can constitute a social word of their own; each of them can also belong to other social worlds. CHWs/CTBAs are not only related to the CB-TB program; in addition, they are parents, farmers, business people, and members of youth, church, or business groups. Each of them is part of other social worlds that are comprised of complicated relationships and interactions among not only humans, but also nonhuman elements. Actors of these different social worlds include collectives or individuals, such as family members; villagers, such as the chief or assistant chief, the elderly, CHC members, TB and other patients, patients' caregivers, colleagues, or neighbors; and facility staff, which includes CHEWs, nurses, clinicians, patient supporters, and staff of the CDC. Actants/nonhuman elements include organizations, both governmental and NGOs; health programs, such as TB, HIV/AIDS, and immunizations; and the places where CHWs/CTBAs can be present and interact with other actors, such as their homes, the health facility, the hospital, their gardens, the market, the church, or the baraza (an important meeting place for the whole village). These actors/actants are not only related to CHWs/CTBAs, but also interact with each other. These social worlds do not exist independently; they are closely related to each other and collaborate in many circumstances in order to undertake cooperative work. Figure 4.1 depicts their relationship with other actors, institutions, and actants in a situational map.

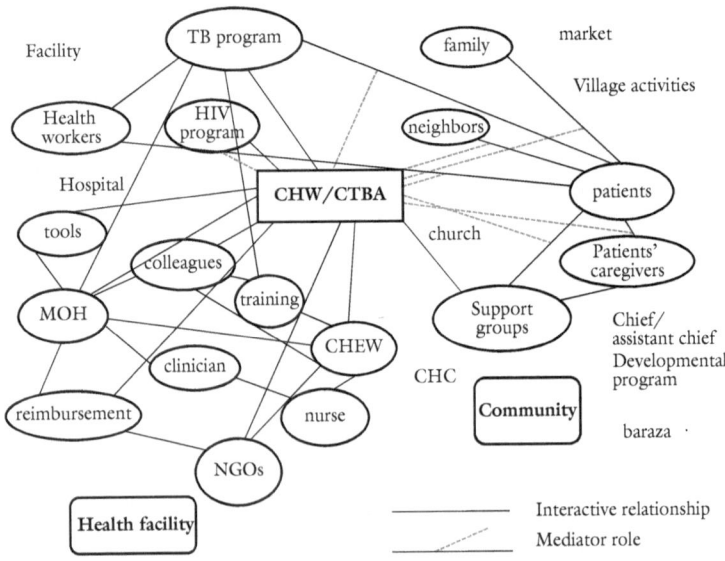

**Figure 4.1.** Example of a situational map (by Phuong Nguyen Thi Mai).

This study's main focus was the social world of the health facility and the community in Kogelo. To a certain extent, the health facility can be understood as a "production world" (Frost, Reich, and Fujisaki 2002) and its "product" is health services; the community might be known as a "communal world" (Frost, Reich, and Fujisaki 2002), in which people live together with shared values and behaviors. In the case of community-based approaches such as the CB-TB program in Kogelo, these social worlds come together for the common purpose of reducing the burden of the disease. The social world of the health facility aims to raise awareness about TB, make accurate diagnoses, provide enough drugs to treat patients, and follow-up with every TB patient. Health workers can meet their patients only when they come to the facility and use standardized protocols and policies to do so. The social world of the community relies on different ways to treat their sick people and struggles with poverty; objection to TB treatment is common. In their work, CHWs/CTBAs in Kogelo have close relationships with both people in their villages and staff in the health facility; they can offer solutions for the problems of both worlds. CHWs/CTBAs can be considered as boundary objects, linking the social world of the health facility and the community. In doing so, they struggle with difficulties arising from both sides as discussed above.

CHWs and health workers in Kogelo appreciate the CHWs close relationships with both people in their villages and staff members in the health facility. CHWs trace defaulters and guide them back to the facility, while health workers equip CHWs with necessary knowledge and support them any time they need help in the facility. "Friendly" is the word CHWs used to describe the nature of their relationship with health workers in the Kogelo Dispensary; they show respect to each other. Recognition and encouragement from the health facility can motivate the CHWs to do their jobs well; if absent, CHWs become discouraged, which may negatively affect their work and passion for their role.

*If we are sick; we don't have to line up, the clinician will give us medicine, because they know that we are busy back there in the villages. (CHW/CTBA 02—Recording 120611_001)*

*And also we praise them when they do something good. You see the way I interact with people, I don't take myself too serious, just work with them as my sisters, brothers, take away the feeling that you're leading board and you encourage them, it can motivate them. (CHEW—Recording 120615_001)*

The relationship among CHWs is of great importance for them to function well. The atmosphere during meetings or discussions demonstrated how friendly and supportive the CHWs are with one another. Most of them mentioned their colleagues in other village as their first contact when they have difficulty convincing patients to get tested or when they need support. The experience of older CHWs is considered helpful to younger workers. In order to overcome difficulties, all CHWs/CTBAs recognize the importance of solidarity and unity among them.

> *We do call each other, then we come here, we share our experience, the difficulties we see in the field, we come together and share here, then we see how to come up with the solution. (CHW/CTBA 07—Recording 120613_002—Translator 01)*

On the other hand, CHWs help provide appropriate information to the community regarding TB. They try to convince community members to go to the health facility for check-ups and assist them with follow-up treatment. Consequently, a CHW's passion does not only depend on relationships with facility staff, but also on their position within their communities. Bhattacharyya mentioned age, gender, ethnicity, and even economic status of CHWs as major factors affecting communities' perceptions of CHWs (Bhattacharyya et al. 2001). However, in Kogelo, the most important characteristic of CHWs identified by the villagers is their ability to keep secrets and protect people's privacy, especially when they work with sensitive issues such as TB or HIV/AIDS. A CHW's reputations regarding work performance plays an important role in a patient's decision of whether or not to trust that CHW. Comparing the results from interviews with CHWs to those of interviews with health workers and TB patients, as well as information from observational notes, indicates a difference of opinion regarding the work of CHWs/CTBAs in Kogelo. These differences include CHWs' ability and effort to work effectively, to acquire new knowledge from trainings, and to communicate with patients in the community. Sometimes the efforts of a particular CHW are recognized and are truly appreciated by both health workers and TB patients. In other cases, the health workers and/or patients do not validate the CHW's own work assessment. Some patients complained about the CHW in their village and would rather see another CHW in a different village; this is mainly due to the fact that they could not find their CHW when needed. Some CHWs were seen as not being qualified to do the job, either because they were not dedicated enough or had limited ability to acquire new knowledge, sometimes due to illiteracy. This reputation can be influenced through community members' own experiences or through rumors and gossip. Also, if community members feel the CHW is someone just like them with no special skills, they do not see the reason to take their advice seriously.

CHWs are aware of their hybrid position as a link between the community and the facility. Some feel closer to the community to which they belong;

others think that they have to be flexible in communication with both sides and keep it balanced.

> We're in the middle, we're mediators. (CHW/CTBA 01'—Recording 120608_003)

> *She's closer to the community but the facility is the reporting point. (CHW/CTBA 01—Recording 120609_001—Translator 01)*

Playing many roles at the same time is a common characteristic of CHWs in Kogelo. As a result, the distinction between "insiders" and "outsiders" in each of the two previously mentioned social worlds involved becomes blurred. Some CHWs mentioned that they can understand and sympathize with patients' conditions, not only because they live in the same village and share the same cultural backgrounds, but also because the CHWs themselves have been through similar situations of illness and difficulty. Some CHWs are HIV positive or have family members living with HIV; some have a husband or relative infected with TB. As such, they use their knowledge acquired through formal training in combination with their real-life experiences to talk to patients about TB. In addition, they are active members of support groups for people living with HIV, for couples in dispute, and can use their own life stories to share with these people.

> *She's very free back at home with the community members. In the community, people know that she's HIV positive, she talks to them about it and how she goes through it … even she shows the patient the drugs she's taking, so she's really convinced a lot of people. (CHW/CTBA 05'—Recording 120613_001—Translator 01)*

The CHW's position as a boundary object is not one of being solely in the middle; they must be flexible to fit themselves in different situations. Their relationship with both worlds is not a simple, unidirectional relationship. Each actor/actant can have either positive or negative effects on the other, thus affecting the potential outcome of the CB-TB program as a whole. Therefore, those involved in the program should be made aware of the CHWs position within the program; the CHWs' work and the success of the CB-TB program are shaped by complicated relationships with other social worlds.

## Discussion and Conclusion

Using the "social worlds" and "boundary objects" framework, the results of this study can be used to discuss and explore if the conceptual framework can help analyze the involvement and position of CHWs in the CB-TB program or whether there is an alternative way of explaining the situation. The framework not only helps expand the analysis from a map of an individual social world (that of CHWs) to their interactions in larger social arenas (the dispensary, the village), but it also allows for deeper understanding of the positions of different actors, boundary objects, and their relations within each arena. Furthermore, suggestions for adjusting the CB-TB program in the future can be proposed from the perspectives of the people who are directly involved in the program: the CHWs/CTBAs, health workers, and TB patients. The results above show that CHWs are the key to the success of a community-based health program; they can effectively link different social worlds, which would otherwise have difficulty connecting and collaborating with each other, together. The social world of professional activity at the health facility has its own universal and standardized approaches to TB control. Nevertheless, CHWs can help to link this "alien" professional world to the local worlds present in Kogelo.

CHW's role as a linchpin between the health facility and the community is not an easy one, as many difficulties arise from both worlds. Programs must do everything possible to strengthen and support these relationships, such as providing visible results that help the community recognize CHWs' work. Being transparent about the program, in terms of what the CHWs can contribute and receive from participating in it, would ensure that there is no misunderstanding regarding CHWs' role among the community. CHWs need to feel that they are part of the health system through appreciation of their work, supportive supervision, solidarity, and appropriate training. These and other nonmonetary incentives are critical to the success of any CHW program. However, questions concerning how long CHWs can keep devoting themselves to their work without receiving any support to alleviate their own poverty, or how to balance CHWs' contribution to the welfare of their community versus their own and their families' welfare, also need to be addressed. Since the CHW's work is related to both the health facility and the community, it should be evaluated and supervised from both sides. Program planners should take into consideration the formal evaluation from the health centers, as well as the informal evaluation, in the form of trust and appreciation from the community. In countries such as Kenya, where coinfection with TB and HIV is high, integration of TB and HIV programs is crucial (Gabriel 2011); CHWs could play an important role as boundary objects, linking and mediating different social worlds. These questions should underlie any initiative involving CHWs and might be a fruitful field for future studies.

How should community-based programs be tailored and adapted to specific situations? The results imply that there cannot be one uniform description for the ideal CHW, nor can there be one standardized CB-TB program that fits every situation. To make such an "alien" world function in the local context, CHWs' connective activities are crucial; however, how exactly they should function depends on how they are perceived in both worlds. As the social worlds of community members are local and specific in nature, differing in place and over time, requirements of CHWs to successfully connect the community with standardized disease control programs and health practices also differ in place and time. As such, policy makers, planners, and managers should be aware of the connective nature of the role of the CHW, understand their social status in different social worlds, involve them in program planning, and support them accordingly.

## Abbreviations

CB-TB:  Community-Based TB program
CHC:  Community Health Committee
CHEW:  Community Health Extension Worker
CHW:  Community Health Worker
CTBA:  Community Tuberculosis Ambassador
MOH:  Ministry of Health
NGOs:  Non-governmental Organizations
NTP:  National Tuberculosis Program
NLTP:  National Leprosy and Tuberculosis Program
TB:  Tuberculosis
WHO:  World Health Organization

## Acknowledgements

We are deeply grateful for the time and insights provided by the participants of the study and to various people in Kogelo, Kenya, for offering their help and assistance during the data collection.

## References

Ayisi, J. G., A. H. Hoog, J. A. Agaya, W. Mchembere, P. O. Nyamthimba, O. Muhenje and B. J. Marston 2011. Care Seeking and Attitudes toward Treatment Compliance by Newly Enrolled Tuberculosis Compliance by Newly Enrolled Tuberculosis Patients in the District Treatment Program in Rural Western Kenya: A Qualitative Study." *BMC Public Health 2011* 11: 515.

Bhattacharyya, K., P. Winch, K. LeBan, and A. M. Tien. 2001. *Community Health Worker Incentives.* Arlington, Virginia: The Basic Support for Institutionalizing Child Survival Project (BASICS II) for the United States Agency for International Development.

Clarke, A., and S. L. Star. 2008. "The Social World Framework: A Theory/ Methods Package." In *The Handbook of Science and Technology Studies, Third Edition*, edited by E. J. Hackett, O. Amsterdamska, M. Lynch, and J. Wajcman, 113–131. Cambridge MA: MIT Press.

Frost, L., M. R. Reich, and T. Fujisaki. 2002. "A Partnership for Ivermectin: Social Worlds and Boundary Objects." In *Public-Private Partnerships for Public Health*, edited by M. R. Reich, 87–109. Cambridge: Harvard University Press.

Gabriel, A. P. 2011. "Evaluation of Task Shifting in Community-Based DOTS Program as an Effective Control Strategy for Tuberculosis." *The Scientific World JOURNAL (2011)* 11: 2178–2186.

Maher, D. 2003. "The Role of the Community in the Control of Tuberculosis." *Tuberculosis (2003)* 83: 177–182.

MOH. 2007. *Reversing the Trends in Health and Development Status: Community Strategy Implementation Guidelines for Managers of Essential Package of Care at the Community Level.* Kenya's Ministry of Health.

Ndege, S., and K. Gitau. 2007. *Community-Based DOTS: Report from a Rapid Assessment in Kenya.* Submitted to the U.S. Agency for International Development by the MSH/SPS Regional Office, Nairobi, Rational Pharmaceutical Management Plus Program. Arlington, VA: Management Sciences for Health.

Niganda, B., J. Wang'ombe, and K. A. Floyd. 2003. "Cost and Cost-Effectiveness of Increased Community and Primary Care Facility Involvement in Tuberculosis Care in Machakos District, Kenya." *International Journal of Tuberculosis and Lung Disease* 7.

Ogumu, I. O. 2011. "Performance of Community Health Workers in Improving TB Management in Siaya District." Master thesis in Community Health and Development, Great Lake University of Kisumu.

Singh, P., et al. 2011. *One Million Community Health Workers: Technical Task Force Report.* The Earth Institute Columbia University.

Star, S. L, and J. R. Griesemer (1989). "Institutional Ecology, 'Translations' and Boundary Objects: Amateurs and Professionals in Berkeley's Museum of Vertebrate Zoology, 1907–1939." *Social Studies of Science* 19: 387–420.

WHO. 2000. "Community TB care in Africa." A collaborative project coordinated by WHO in Harere, Zimbabwe.

WHO. 2010. *Global Tuberculosis Control.* World Health Organization.

CHAPTER 5

# The Forgotten Age Group: Why Children Ages 5 through 14 Are Succumbing to Malaria. An Exploratory Study in Western Kenya

*Marianne Eelens and Agnes Meershoek*

## Battling a Historic Disease

The first accounts of malaria were given by Hippocrates (460 BC–370 BC), granting him the recognition as the first physician to clearly describe the different forms of the disease. This Greek physician made the accurate link between ecological conditions and disease, while identifying fever as a key initial symptom (Cunha and Cunha 2008). Malaria is a preventable and treatable disease; however, it remains one of the most devastating global health problems of our time. An estimated 207 million cases of malaria occurred in 2012, of which 627,000 resulted in fatalities—most of them being children under the age of five in sub-Saharan Africa. To address the vulnerability of this population, the sixth United Nations Millennium Development Goal (MDG) was established in 2000 to guide funding and malaria efforts toward this population (WHO 2013). Specifically, MDG 6C aims to halt malaria and begin reversing the incidence by 2015 through successful partnerships combining effective prevention and treatment methods (UN 2011).

The World Health Organization (WHO) recommends vector control interventions, preventive therapies, prompt parasitological diagnoses in all suspected malaria cases, quality assured treatment, and strong malaria surveillance as keys to controlling and eliminating malaria. Recommended prevention methods are larval source management, indoor residual spraying, and insecticide-treated nets (ITNs). Indoor residual spraying in homes and sleeping areas where the endophilic vector is prevalent are effective approaches to reduce malaria vectors (WHO 2013). ITNs are also highly recommended as a vital preventive method; they create a physical barrier between the female *Anopheles* mosquito (vector) and humans (hosts) (Division of Malaria Control (DOMC) 2009). Most mosquitoes bite at night; the netting material is treated with insecticide, making ITNs one of the most effective tools in our armamentarium against malaria infection. ITNs have been shown to reduce malaria transmission by up to 70–90 percent in an efficacious and cost-effective manner (Gatei et al. 2010). However, providing universal coverage and making ITNs available and accessible to all of those living in endemic regions remains a challenge. In addition, the quality, durability, and proper use of ITNs have proven to be additional barriers to successful coverage levels and malaria prevention (Chuma et al. 2010a). As the sustainability and accessibility of prevention methods in resource-poor settings remain problematic, treatment continues to be a crucial priority in battling malaria-related deaths.

In 2006, the World Health Organization (WHO) provided a set of guidelines for the treatment of malaria; these guidelines take into account local drug resistance and the limited capacity of the public health sector. Before the initiation of any treatment, the WHO (2013) suggests prompt diagnosis of malaria by confirming parasites in the blood through microscopy or by use of a rapid diagnostic test. However, due to the unavailability of diagnostic tools in many local health centers, malaria is often overdiagnosed, resulting in unnecessary treatments (Whitty et al. 2008). For this reason, the Kenyan government accepted global recommendations and recently adopted a policy of universal diagnosis in which rapid diagnostic tests are utilized to diagnose malaria. This method is a simple, effective solution that addresses misdiagnosis and consequently reduces unnecessary treatment and expenditures (USAID 2013).

The recommended treatment for uncomplicated *P. falciparum* malaria is artemisinin-based combination therapy (ACT) (WHO 2013). The advantage of artemisinin is that it clears parasites much more quickly from the bloodstream than other previously recommended antimalarials, such as quinine or chloroquine (Jambou, Le Bras, and Randrianarivelojosia 2010; Gill and Beeching 2004). Although effective, access to ACTs is often hindered by the limited availability of the drug in the retail sector due to its high price, popular beliefs about the cause of malaria, and treatment seeking behavior, as well as inaccessible health facilities and dispensaries (Patouillard, Hanson, and Goodman 2010; Chibwana et al. 2009). This results in malaria patients resorting to self-medication and traditional medicine, characterized by delayed and inappropriate treatment of malaria with an increased mortality risk (Mungai 2009).

However, vast improvement can be seen in the control of malaria. Between 2000 and 2010, the incidence rates of malaria have decreased by 17 percent. In addition, since 2000, the global malaria mortality rate has fallen by 25 percent. In the same decade, ITN use by children under the age of five in Sub-Saharan Africa increased from 2 percent to 39 percent (UN 2012). Initiatives such as the MDGs have played a major role in raising both funding and global attention concerning this disease. International disbursements for malaria significantly increased from $100 million in 2000 to an estimated $1.97 billion in 2013, enabling endemic countries to increasingly provide prevention, treatment, and diagnostic services. However, in the past few years, this has plateaued and remains short of the projected minimum of the $5.1 billion needed per year between 2011 and 2020 for achieving universal coverage of malaria efforts (WHO 2013).

An additional concern is the fact that malaria prevalence in some endemic areas is actually increasing among certain populations. In many malaria endemic countries in Sub-Saharan Africa, the main focus is on those most vulnerable to contracting the disease—pregnant women and children under the age of five. These populations are specifically targeted and educated on malaria during clinic visits while pregnant, through vaccination campaigns, and during other such contact instances (DOMC 2009; Webster et al. 2005). In Kenya, these efforts have proven to be successful, as ITN use for children under the age of five increased from 3 percent to 47 percent between 2000 and 2010 (UN 2011). As various studies report similar success stories in ITN use among children under five years of age, one cannot help but wonder what happens to those who just miss this age limit during periodic, mass campaigns (Chuma et al. 2010a). From a global health perspective, this question is highly relevant, as a staggering 80 percent of all human-to-mosquito transmissions occur in individuals above the age of five, with most of these occurring in older children (Noor et al. 2009). The 2010 Kenya Malaria Indicator Survey indicated that while the latter population is less vulnerable to the detrimental effects of malaria, their undeveloped immunity to parasites does make them a dangerous reservoir of asymptomatic infections. With thirteen percent of five- through fourteen-year-olds testing positive for malaria in Kenya, compared to five to ten percent of children under five years of age, increasing attention should be placed upon the inclusion of older children in malaria control (DOMC et al. 2011).

## Uncovering the Hidden Truth

Most scientific literature focuses on the vulnerability of children under five years old; this age group is placed on the frontline by initiatives such as the

MDGs. But what is actually happening backstage? And what implications do WHO's standard recommendations on prevention and treatment actually have for malaria risk in local settings? We conducted a case study in Western Kenya to signal possible shortcomings in the commonly-assumed approach focusing on children under five and explored how this relates to malaria prevention and care in children ages five through fourteen. Obtaining an improved understanding of the effects of the inattention placed on this age group, while looking at specific social aspects, were key objectives of this study. In this context, one aspect that was extensively considered during this study was ITNs—regarded as one of the most effective malaria prevention innovations (Gatei et al. 2010). How gender roles affect ITN use in the household and the manner in which treatment is sought were also considered. To truly understand the figures on ITN use and treatment-seeking behavior, one should consider whether these technologies are actually reaching their true potential. To do so, Leach and Scoones suggest a "slow race" approach, which emphasizes the fact that technological fixes present an insufficient solution to health and development problems; this approach also stresses that social, cultural, and institutional dimensions are key factors to be taken into consideration when tackling these problems (2008, 124). Here, a one-size-fits-all approach is not the solution; a bottom-up, participatory approach placing individuals at the center is key for providing successful, context-specific solutions. The qualitative approach used in this study creates a crucial understanding of social and cultural factors that impact how and why ITNs and certain treatments are accepted or rejected. Subsequently, even if standardized methods recommended by the WHO such as ITNs and ACT treatment are readily available, the remaining social and cultural barriers that are associated with the disuse of such technologies need to be understood in order to optimize results.

The rainfall, humidity, and temperature in the areas surrounding Lake Victoria in Western Kenya have created perfect conditions for the *Anopheles* mosquito to thrive throughout the year. For that reason, this endemic area was chosen for this study and included the districts of Kisumu East, Kisumu West, Rachuonyo (North and South), and Nyando in the Nyanza Province. A total of 12 focus group discussions (FGDs) with 138 female caretakers were conducted in these locations. FGDs are seen as a quick and convenient manner to collect data from numerous participants simultaneously. In addition, FGDs are a suitable way to use group interaction to find out more about participants' knowledge and perceptions (Kitzinger 1995). Female caretakers were selected; they remain the principal caretakers of children in the household, enabling them to provide a detailed account on the manners in which malaria is prevented and treated in older children. Participants were selected on the basis that they are female caretakers who care for at least one child between the ages of five and fourteen. Notwithstanding a recent policy adoption promoting universal diagnosis of malaria (USAID 2013), it is crucial to understand what enables or disables the prevention and treatment of malaria in older children. This study focused on ITNs and treatment as they form an integral part of Kenya's National Malaria Strategy, which was based on WHO guidelines and aims to achieve universal coverage for all at-risk populations (DOMC 2009). In addition to these topics, the discussions, which were conducted in English but translated to Dholuo (local language spoken by Luo people), also touched upon the relationship between gender roles and decision making regarding ITN use.

## The Inverse Relationship of Age and Treatment Priority

Female caretakers revealed that an abundance of diseases and ailments are debilitating their children's daily lives, but malaria is causing the greatest burden. On average, children ages five through fourteen suffer from malaria an estimated nine times a year; roughly a quarter of the women reported up to twelve episodes in a year. Symptoms such as fever, shivering, vomiting, stomachache, headache, loss of appetite, red eyes, general body weakness, weak and painful joints, and having a cold caused female caretakers to suspect a case of malaria. In addition, a child having a runny or bloody nose, yellowish skin, diarrhea, pimples, weight loss, dizziness, convulsions, and changes in lip color were also commonly connected to malaria. Interestingly, although typhoid fever shares very similar symptoms to malaria, the connection between flu-like symptoms and typhoid fever was rarely made. When a child falls ill, flu-like symptoms were always linked to malaria. Rural households in different parts of Kenya, where malaria is endemic, often associated fever with malaria (Chuma et al. 2010b). Moreover, most women understood the etiology and harm the disease can afflict, but they were often unable to distinguish symptoms from a true diseased state. A study conducted in rural Western Kenya indicates similar results wherein flu-like symptoms are labeled as diseases in their own right (Ruebush et al. 1995). On one hand, as malaria is prevalent in Western Kenya, it may be an obvious response for preoccupied mothers to view any symptom as a disease. On the other hand, this can lead to overdiagnosis and unnecessary treatments. The reported frequency of malaria in children ages 5 through 14 in this study might therefore be an overestimation.

Female caretakers were well aware that if a child is suspected of having malaria, he or she should go to the hospital for proper treatment. Ideally, the child is given the commonly used ACT medication *Coartem* at no cost or time delay. However, reality proves different. Numerous issues can arise that prevent the sick child from being taken to the hospital or treated adequately. A major issue is the constant existence of financial constraints.

*Because of money, we can't take them to the hospital directly; we maybe give them painkillers just to forget about it (FGD 10, 23 June 2011).*

If the disease persisted, the child was taken to the hospital. The issue of an empty wallet does not just translate into a lack of treatment; it spirals into a much deeper and complex problem in which destitution plays a central role. One of the main aspects directly related to this problem is the limited amount of food available in many households. Even if caretakers could access proper treatment, lack of food stands in the way of the timely intake of treatment and adherence to treatment regimes. The mutual combination of financial constraints and lack of food made some respondents anxious about taking malaria medication on an empty stomach. Some believed that taking medication on an empty stomach can be poisonous, prompting them to resort to "safer" but less effective alternatives.

*The economic crisis is a very big problem. Because it may come at a time they[1] are given the medicine to take, but when you take the medicine they don't have anything in the stomach. At the end of it all it becomes a poison. There's a mother somewhere in the village here who died because she was taking the medication and did not have sufficient food to eat. So you realize that because there is no food, there is a lot of poverty, the people concentrate on the food issue and not the health matters. So it becomes a problem (FGD 11, 27 June 2011).*

It is not only small households that cannot treat malaria properly due to financial barriers; many governmental health institutions faced similar challenges. Initial government promises of providing malaria treatment at no cost were not always kept, fueling uncertainty and anger among the female caretakers and children.

*At some point the government promised towards the people that malaria treatment is free. So the mothers were going and going. Then it came a time that now, it is charged, they were told it is free and they go to the hospital they told it is charged. So you find you may be coming from home knowing that you're going to get free treatment, you don't have any money. So going there you find that a lot of money you need to pay. You have to come back, and come back, there comes the problem (FGD 1, 25 May 2011).*

One mother mentioned that her child did not survive the seven kilometer walk to the nearest health center—dying in her arms underneath the blazing sun. The distance to the nearest health facility plays an important role in seeking appropriate treatment. According to the women, young age and the severity of the child's condition are deciding factors as to whether the mother seeks treatment immediately or waits a day or two. Women had contradicting views about seeking treatment for children under the age of five versus those ages 5 through 14. Some women stressed the fact that they treat all of their children equally and seek treatment whenever they suspect malaria, no matter the age. Others reported a difference as to when treatment is sought. There are several reasons why the treatment-seeking behaviors of caretakers can differ for children under the age of 5 and older children. As mentioned earlier, a major challenge to seeking (timely) treatment is the distance to the nearest health facility. The size and weight of the child is a deciding factor in determining whether it is necessary to take a *boda-boda* (bicycle taxi) or *piki-piki* (motorbike taxi) to a distant health facility; this influences when treatment is sought.

> *So they would urgently go with the young ones because they are saying that the older ones you must get enough finances to take them to the hospital. And these ones [small children] you can just carry them (FGD 12, 27 June 2011).*

As financial constraints were constantly present, mothers often resorted to herbal medicine, or over-the-counter medicine found nearby, to treat the older child at home. That said, mothers felt more comfortable treating the older children at home compared to children under the age of five; the older children were perceived as being physically stronger. Older children living in malaria-endemic regions in Western Kenya are likely to have acquired some immunity to malaria through natural and continued exposure and could, therefore, be better suited to fight recurring disease than their younger siblings (RBM 2008).

In addition, older children can better express their pain and concerns. Immediate action was taken with younger children; the extent of suffering was often not understood. The difficulty of establishing the cause of the young child's symptoms made it more challenging and worrisome for the mother. Therefore, due to the scarcity of resources, caretakers have relentlessly searched for solutions and are creating alternatives for treating malaria in older children (Chuma, Okungu, and Molyneux 2010). Another reason why treatment seeking for the two populations differed is due to the fact that the level of attachment between mother and child varied with the child's age.

> *First because the young one you are close to and you still have so much attachment to the baby, so she would first rush to treat the young one, then maybe the older ones maybe rush for a painkiller then later seek treatment and take them to the hospital (FGD 7, 17 June 2011).*

About a quarter of the female caretakers in the FGDs were widowed. Widowhood often resulted in more lenient measures for treating malaria in older children; the mothers struggled to run a large-sized household single-handedly. Close observation of a mother's role in society revealed that she is expected to fulfill numerous tasks during the day, which include putting food on the table, fetching water, mending torn school uniforms, cleaning the house, working in the *shamba* (food garden), and so on. These widows were found to be both poor and extremely busy, which can be associated with higher poverty and mortality rates than other groups (Ntozi 1997). This situation made it much more difficult to care for a child that requires costly treatment; the widow had very limited finances and time available. Therefore, widows were often forced to delay treatment in older children, whereas treatment for children under five years old is sought sooner as it was deemed more urgent. Supporting evidence for this finding was not found, but as widowhood is common in Sub-Saharan Africa, this matter deserves more attention.

The distribution of ACTs by community volunteers allows for home management of malaria, a strategy focused on providing treatment to children under five years old and implemented to avoid the use of ineffective, over-the-counter drugs provided by the unregulated private sector (Health Department 2010). Providing the same care for older children would be beneficial; these children are more often kept at home during illness, as it is too costly for them to be taken to distant facilities—a common affordability issue also reported in other literature (Chuma, Okungu, and Molyneux 2010).

The cost of treatment was a major deciding factor that influences the way treatment is sought for children of different ages. ACT costs varied in the different locations that were visited. In theory, treatment for children under five years old is free, while children above the age of five have to pay 20 KES for treatment (at the time of the study). In other locations prices are different:

*Coartem costs 200 shillings for children while for adults it costs 600 shillings; it's too expensive for us (FGD 10, 23 June 2011).*

Having to pay more to treat a child above the age of five created an obvious disadvantage and increased the chances of caretakers resorting to less effective treatment regimes.

A study conducted by Chuma, Okungu, and Molyneux (2010) in four districts of Western Kenya revealed that more children under five (29.3 percent) were treated in public health facilities, in comparison to older children and adults (19.2 percent). As a result, older children were more often treated with drugs bought in the informal sector (shops and chemists) or with herbal medicine. This evidence coincides with the findings in this study, in which older children were often treated with painkillers (such as *Panadol*) or herbs such as *dwele* and *mwarobaine*. As the poorest populations are most affected by malaria, it can be understood why caretakers prioritized treatment for those most vulnerable in the family (Chuma, Okungu, and Patouillard 2010c). A higher percentage of older children were not treated for their illness due to financial constraints. Ironically, the above-mentioned factors that are disadvantageous to the health of older children are ameliorated by the precautionary measures, such as insecticide-treated nets, that mothers took to prevent malaria in their youngest children.

## Net Etiquette: The Most Common Practices within Households

Similar prioritization measures were used for the prevention of malaria. While mosquito repellents and other preventive measures were available in stores, limited finances made anti-mosquito products inaccessible to many households. Therefore, caretakers were creatively dealing with the situation by attempting to chase, kill, or avoid mosquitoes. Several actions were taken in- and outside the home.

*They use cow dung, and then they dry one, they burn, and it has some bad scent and the smoke it sends away the mosquitoes (FGD 12, 27 June 2011).*

Caretakers had been encouraged by community health workers and other malaria campaigners to perform various actions around the home such as:

*Clearing the compound, the bushes around, and also draining stagnant waters (FGD 4, 1 June 2011).*

*She ensures that her house is well-ventilated so that now in the night the mosquitoes would not get in (FGD 3, 1 June 2011).*

The latter two measures do not hold true for most female caretakers living in Nyallenda B—part of the Kisumu slums. Here, they lived in small rental homes and did not feel responsible for taking care of the surrounding properties. More modern methods such as ITNs and burning of mosquito coils were also used. Unfortunately, despite ITNs providing great protection against malaria, there were many issues associated with their ownership and usage. There were far too many households struggling to protect everyone during the night.

*They are giving you but they are considering you and the baby. But if they could be given the nets considering the number of children to be protected from malaria [that would be better] (FGD 2, 26 May 2011).*

The problem is that ITNs become an item poor families are urged to purchase with their own resources—resources they often do not have. In addition, as ITNs age, their insecticide status diminishes and requires investment in retreatment. As a result, bed nets (treated or untreated) become an item of luxury, purchased when perceptively needed.

*They just get when they feel the need of having a net when they start biting, mosquitoes, or the frequency of malaria infection in your house [increases]. I should have a net, then they go and purchase at the shop by themselves without any aid from the ministry or hospital (FGD 5, 6 June 2011).*

*Before they were given they had nothing, they prefer that better to have some ugali [local dish] than having a net, than buying a net at a very expensive price. They were taking painkillers to keep them going" (FGD 10, 23 June 2011).*

Women had an understanding of the importance of bed nets and clearly linked the lack of net protection with mosquito bites and, consequently, malaria. But as scarcity of bed nets is common, women preferred to cover the youngest child first; older children were seen as better capable of putting up both a physical and physiological fight.

*The new nets you give to the kids, the young ones. And the older ones they give the ones that are a bit old, because they can at least also struggle and fight with the mosquito (FGD 9, 22 June 2011).*

Research across Sub-Saharan Africa also showed similar results in which the best quality nets were used to protect the youngest children (Githinji et al. 2010; Noor et al. 2009; Eisele et al. 2009; Tsuang, Lines, and Hanson 2010; Baume and Marin 2007). Although female caretakers perceived older children as having stronger immune systems than their younger siblings, opposing evidence suggested that this is not the case. When small children sleep under a net, their body is unlikely to acquire high immunity to infection. When the child's age increases and the net is passed on to a younger sibling, the risk of malaria infection is also increased (Bejon et al. 2009). However, long-term studies indicated that nets are an effective tool in reducing malaria-related morbidity and mortality (Lindblade et al. 2004; Binka et al. 2002; Diallo et al. 2004).

With increasing age comes increased responsibility—an aspect also true for malaria prevention and the use of ITNs. Older children often slept in the common room or kitchen on mats where their nets had to be retied each evening, a task often placed in their hands. As these children may stay up later than their parents, the lack of supervision allowed the children to resist or refuse net use.

*Yeah like my boys, they do complain. They say "mommy today we are not spreading the net, because I don't even feel the mosquitoes biting us." So sometimes I just force them to spread the net, sometimes I just sleep when I'm tired I just go to bed early and leave them and they just sleep like that [without a net] (FGD 10, 23 June 2011).*

Small children usually slept in the bedroom with their parents. Sleeping on a raised bed and mattress made the use of a net much easier. Older children mostly slept on reed mats on the floor, making it difficult to extend the net all the way down to the floor to ensure full coverage. In addition, nets hung above sharp-edged reed mats were damaged more easily than those used by small children sleeping in beds.

*The ones [nets] that are laid on the mats they get torn and they don't survive for the 5 years that is usually said and because of the resources and poverty, you may find that they now use those just to survive for the sake of (FGD 3, 1 June 2011).*

A previous study supports this finding, mentioning that nets given to children under five years old were of considerably greater quality than those used for older children. By use of a mean hole index, this study indicated that nets for children under five had the fewest holes (5.5 holes); those used by older children had the most (13.0 holes) (Tsuang, Lines, and Hanson 2010).

Sleeping in the common room or kitchen also affected the quality of the net. Nets that were hung above mats often come in contact with the soil. In addition, depending on the size of the home, either the common room or kitchen was used for both cooking and sleeping. When nets were hung in these rooms, they can obtain unpleasant cooking odors or become dirty at the bottom, prompting women to wash them. Without money to buy insecticide to re-treat the net, it was partly stripped of its protective factor and effectiveness.

Scarcity forced some to sleep without an ITN during the year-round presence of mosquitoes and can create jealousy among older children.

*There is insufficiency where the older ones see that the most preference is given to the younger ones: "You [young child] are given nets and we [older children] are not given" so they are jealous because the priority and resources are concentrating on them [points to baby] (FGD 11, 27 June 2011).*

This sense of jealousy can be worsened when the older children see older nets around the home being used for other purposes than their protection. The expected results during malaria control campaigns are frequently not produced when women continue to use bed nets in much more creative ways than the product's original intent. This creates a disparity in net ownership and actual use (Dunn, Le Mare, and Makungu 2010). Fishing is one alternative bed net use found in this study. About 84.5 percent of nets used for fishing along Lake Victoria were received for free or at subsidized prices (Minawaka et al. 2008). Nets can also serve other purposes such as mops, dusters to clean, sponges to wash their bodies, or places to dry fish. Walking in the Nyallenda slums exhibited a large number of alternative uses. Many fences around homes or gardens were made of bed nets; even the walls of community pit toilets consisted of this netting material. In some places, old nets were simply stored in cupboards for possible future reimbursements:

> *The government wanted the old nets … and they give replacements, for new ones (FGD 4, 1 June 2011).*

As this reportedly occurred once in a particular village, these caretakers now kept their old nets in storage with hopes of benefiting from a similar act in the near future. Upon receiving new nets, some caretakers immediately replaced them with those that are in current use. Others preferred to store new nets away; they were keen to stretch the use of the older nets way beyond their protective lifespan. In 2003, Alaii et al. reported that about 30 percent of the nets received for free were left unused in Western Kenya.

This study has illustrated that bed nets are not simply present in communities for creating a barrier between mosquitoes and humans. Instead, there are many possibilities and decisions to be made regarding their purchase and usage. While keeping in mind that the Kenyan society has a strong tint of male chauvinism, who really has the ultimate say when it comes to net usage inside the home?

### "I am the Boss!"

As stated by the women, the mother is the main decision maker when it comes to malaria prevention in the household. One woman took it even further and proudly said aloud: "*I am the boss!*" (FGD 3, 1 June 2011). Women stated that they are better informed about malaria treatment and prevention than their husbands. This is due to the fact that mothers were exposed to this knowledge when they visited the ANC during pregnancy or when taking a sick child for treatment. Naturally, this finding is limited; men were not included in this study. Other studies argued that men are the main breadwinners and decision makers in the household (Tsuang et al. 2010; Chibwana et al. 2009). Moreover, women were not empowered to make health decisions regarding their children, independent of their husbands (Chibwana et al. 2009). During the FGDs in this study, however, women portrayed themselves as adamant and confident women who take responsibility into their own hands. Informal conversations with males also revealed that women are often better informed on malaria and are therefore the ones in charge of childcare, including the prevention of malaria. However, conversations with men were not structured as with the FGDs, and the number of men interviewed was limited. It could be that women attended health clinics more often during pregnancy, or with an ill child, and were more exposed to malaria campaigns, leading to the women having more decision-making authority when it comes to malaria. In addition, men could be busier earning an income and therefore had little time to care for a sick child (Chibwana et al. 2009).

The difference in knowledge between the parents translated into the husband often being unconcerned about malaria prevention, while the mother was very keen on using a net. A common statement was that if the mother does not hang up the net, her husband would simply sleep without this protective measure. In addition, men were often away from home and unaware of the state of their children's health. Therefore, women were expected to fulfill all of the domestic tasks on their own and were sometimes even criticized for things they did not perform according to the man's wishes.

> *The father will just go: "you say the child is sick, what did you do, what action?" But he will not be there, perhaps he will be coming in the morning, in the evening, and leaving in the morning ... not knowing what's happening (FGD 3, 1 June 2011).*

A husband's lack of knowledge can also make him apprehensive regarding the negative repercussions ITNs can have on his health. According to the women, the fear of suffocation was predominantly present among men; therefore, they preferred to sleep under a net that was not treated with insecticide—a similar fear that was sometimes passed on to their children.

> *Some children who also [as their father] do not want to sleep under a net. That you cover them with nets from downwards [the net is laid on the lower body], especially the younger ones who sleep away from the parents. So you never know what they are going to do at the other side, so they don't even mind having the nets, they don't see the need of having those nets because they claim that it makes them suffocate. So if they will be covered they cover them half-way (FGD 12, 27 June 2011).*

As expected, the difference in knowledge and perspectives concerning malaria prevention has the capability of creating tension between husband and wife. This can easily escalate when a man comes home intoxicated, making it more difficult for the woman to sleep under a net herself.

*Okay there comes a time, you encounter men that have different personalities, some like drunkards, then he comes and you hang your net there decently, go inside to sleep and caring. So you find the hung net drops and then he's like "ey move these things away from my side." It is now causing some inconvenience to him as he sleeps, so there comes a challenge (FGD 6, 6 June 2011).*

Some women resolved this issue cleverly, waiting until the man falls asleep; she then rehung the net and returned to sleep.

*The husband may complain that he doesn't want to sleep under the net, maybe he comes home when he's drunk, he doesn't feel like sleeping under a net, now it will force you to, obviously we are sharing one bed, now it shall force you to move to the other side so he sleeps on the other side (FGD 10, 23 June 2011).*

There were also cases in which husbands insisted that the best, newest nets should be allocated to them; these husbands believed that the children should sleep under older nets or simply sleep without one. This created another point of conflict as the mother was committed to protecting her children. This may either reflect the selfishness or lack of specific knowledge of the husband regarding the vulnerability of children. However, it may also be done to keep the breadwinner in the house malaria-free. Women in Kowidi, a small remote community east of Oyugis, shared their disagreements with their husbands and showed how they attempt to resolve these disagreements:

*With the heart of sympathy that the mother has, let them give the nets they have to the younger ones and them [parents] they remain without net. So it becomes a point of conflict, 'how can they just give the children and leave us without the net [husband says].' This has also become a point of dispute; the man wants the new net and replace with the old one, give the old one to the children. Mosquitoes are so much here that no man would dare [to sleep without a net] ... the men don't get involved so much with the health of the children. So you find the mothers so much concerned, and the man with the ego that the man just wants to protect themselves. Sometimes even in the morning, come back late in the night or in the evening, they [men] don't know the condition of the health of their children, they simply come and want to sleep comfortably under nets (FGD 5, 6 June 2011).*

For men who truly understood the importance of protecting themselves and their children with nets, limited conflicts existed between the husband and wife about this subject. This is due to increasing education and sensitization on the importance of net use. In the case where both parents understood the importance of net use, the main conflict concerning net use between the parents was about who should receive a net first. One woman noted:

> *The only point that would be a disagreement, is the scarcity of nets, is who to sleep under a net and who not to sleep under a net. But not "should we sleep under a net" (FGD 10, 23 June 2011).*

Women have created several solutions to disagreements with their husbands. When a man refused to use a net, she relentlessly tried to convince her husband about its importance for staying malaria-free. If this did not work, women ignored their husbands and let them sleep without a net, while they made sure to cover themselves and the children. This sometimes forced them to sleep separately from their husbands. Several women were fearful themselves of treated nets disrupting one's health—all others were comfortably using the ITNs present in their homes.

> *A child who slept under these treated nets, they were so powerful with such kind of treatment, so the child was taken to the hospital for taking that kind of stuff. And also some pregnant mothers also felt some problem with the use of new nets, so that's the only fear that has been because there's some instruction that it has to be hung for some number of hours because it is very powerful (FGD 11, 27 June 2011).*

An earlier study conducted in Western Kenya revealed that the majority of the participants were fearful of the insecticide smells affixed to new ITNs. Fears of suffocation and other harmful effects were solved by using older nets in which the smell had worn off, or by washing the nets to reduce the insecticide. Participants were not necessarily scared of the harmful effects of the nets; however, fears were instilled in them during a randomized controlled trial. Individuals from control villages spread rumors regarding ITNs in an attempt to outsmart those included in the intervention group who did receive ITNs (Alaii et al. 2003). FGDs conducted in the Imo River basin community in Nigeria, also revealed that some residents believed that ITNs affect the breathing of pregnant women: "If the chemicals can kill mosquitoes instantly, they can also kill people" (Chukwuocha et al. 2010, 120).

## Moving Forward Toward Successful Malaria Control

Ensuring sustainable malaria control is crucial to achieve MDG 6c. Approximately 12 percent of postneonatal child deaths globally and 21.7 percent of postneonatal child deaths in Africa were attributed to malaria in 2010; malaria control is also expected to contribute to MDG 4a (reduce under five year olds' morality rate by two-thirds between 1990 and 2015). MDG 1 (eradicate extreme poverty and hunger), MDG 2 (achieve universal primary education), MDG 3 (promote gender equality and empower women), MDG 5 (improve maternal health), and MDG 8 (develop a global partnership for development) are also expected to benefit from successes in malaria control (WHO 2013). The historical focus on children under five has done a great deal for malaria control in Kenya. However, successes could be compromised if children ages five through fourteen—the main transmitters of the disease— are left unconsidered.

This research suggests that the focus on malaria control for children under five in Kenyan policy is not only in line with international guidelines' focus on children under five, but also coincides with (or even enhances) priorities given to children under five within households. The research also shows how this may enhance the risk of malaria in children between five and fourteen years of age. The research describes how the standard focus on children under five, in line with caregivers' priorities and in combination with social and microeconomic conditions at the household level (considering the inadequacy of currently used nets), actually limits, rather than promotes, protective and curative measures for children under five. To avoid unintended effects of standard policies and guidelines, and in line with Leach and Scoones' "slow race" approach, we therefore plead for a participatory approach in which standard guidelines are translated to fit local conditions by starting from the perspective and experiences of local female caretakers.

It is worth noting that the results produced by this study are not exhaustive, and similar research in other settings is likely to produce different results. Results in this study showed that there are several bottlenecks regarding the control of malaria in older children in Western Kenya. Standardized strategies (such as those recommended by the WHO for children under five) can unfold differently in local settings or when applied to another age group, even to the point that they have reversed effects on the community's health. Due to limited financial resources, combined with the advised focus on children under five, treatment for older children is often sought at a later stage as compared to children under five; herbal remedies or painkillers are common alternatives to regular treatment in these older children. In addition, female caretakers—the main decision makers regarding net use in the household—prioritize nets for children under five, as they have learned to consider them most vulnerable. Adequate use of the bed-nets is impeded by practical issues, such as sleeping on the floor on a mat that needs to be removed during the day. As a result, older children are more likely to sleep under a damaged net or under no net at all, compared to their younger siblings. Finally, bottlenecks related to gender and malaria control were revealed. In this sense, the question becomes whether malaria initiatives, such as the MDGs, have been formulated comprehensively enough to ensure sustainable malaria control, especially as it should also be applied to another important age group.

Producing answers to this question is complex; the formulation of policies and initiatives are dependent upon numerous factors, i.e., economic, cultural, social, and political factors. In addition, producing an overall goal, such as "to halt malaria and to begin reversing the incidents by 2015 through successful partnerships combining effective prevention and treatment methods," is implemented and accepted differently in various regions across the globe. Therefore, having set recommendations and strategies in place to tackle malaria is both necessary and important; however, a thorough understanding of the local setting to ensure success is required. While vital steps toward the success of this goal have been made, causes for concern remain.

One cannot say with certainty that initiatives such as the MDGs are the direct cause for increasing malaria-related problems in older children. The standard response for children under five, especially when the use of impregnated bed nets are involved, may need to be adjusted to the practical constraints a caretaker experiences when she considers the use of similar nets. What we do know is that malaria prevalence in older children in Kenya is on the rise and that these children are often more neglected in terms of care and prevention. This study presents what is happening "backstage" when malaria efforts target children under five. The focus group discussions with female caretakers provide important lessons on the social factors related to malaria control and are essential to tailoring interventions to local settings.

These women are well aware that malaria in a young child is dangerous and potentially detrimental. Children ages five through fourteen are often neglected and deprioritized when it comes to malaria control, both in treatment and net use. Although this is based upon the perspectives of female caretakers (and not policy makers), an improved understanding is obtained regarding the bottlenecks to malaria control in Kenyan households. Treatment-seeking behavior is delayed and often based on alternative medicines, such as painkillers or herbal remedies. ITNs for older children are either not an option or of considerably less quality than required. Therefore, it should come as no surprise that malaria infection is highest among children ages five through fourteen in Kenya (Noor et al. 2009; DOMC et al. 2011). Older children are a particularly important group for two reasons: they make up a large fraction of the total population in African countries, and their immunity against parasites, which regulates the risk of blood stage infection, has not yet developed (Noor et al. 2009). Therefore, if quick action is not taken, the malaria incidents in children ages five through fourteen in Kenya will soon surpass that of children under five years old. In numerous locations, this has already happened (DOMC et al. 2011). While it is important that special attention is paid to the most vulnerable populations, the fact that older children represent the greatest reservoir of infections should also be of great concern (Noor et al. 2009). If the number of deaths in children under five years old has to be reduced (these children currently account for 85 percent of all malaria deaths), those transmitting the disease need to become a much greater focus (Heath Department 2010). Older children may not have been purposively neglected and deemed unimportant by malaria campaigns; the special attention placed on children under five, however, is taking a toll on the health of older children and will hinder malaria control efforts (Winskill et al. 2011). Therefore, quick action is required by creating additional resources; older children run the risk of soon becoming both the greatest disease sufferers and transmitters (Winskill et al. 2011).

It is also important to recognize the impact gender roles can have on the manner in which ITNs are used within the household. As malaria vulnerability is impacted by biological and social factors, it is important that both men and women understand its implications. Therefore, men should be more involved with malaria campaigns so that disagreements can be avoided, and they can provide provisions of care (RBM 2006). In practice, this is easier said than done. Men often leave the home during the day for work or during the night for social activities; women spend their day at home taking care of the children (Dunn et al. 2010; RBM 2006). In some households, these traditional roles create tension when a child falls ill; it is the woman who is expected to fulfill her duty by keeping the children healthy. In other respects, Dunn et al. (2010) found male social activities, such as alcohol consumption, to be criticized, as these activities divert essential funds from malaria prevention and treatment. For these reasons, men should be encouraged to attend prenatal clinics with their pregnant wives in order to create a better understanding between parents about childcare. This sensitization could decrease the fear of suffocation under a net or the disapproval of net use among both men and women. Decreasing the disagreements over ITN use in the household and overcoming bottlenecks that hamper optimum protection of all family members can occur with proper education.

In addition, discussions with female caretakers also revealed that treatment and prevention measures sought for older children are strongly tied to the financial situation and perspectives of these women. Tackling the problem of poverty is an enormous, crucial first step. However, having enough finances to purchase ITNs and treatment does not solve all issues. To reach the full potential of ITNs and malaria treatment, making these products available and accessible to the wider public is key. However, overcoming these barriers requires a deep understanding of the social and cultural dynamics that are present within households. It is only when all of these factors are taken into account that the control of malaria will come full circle and succeed.

## Note

1. All quotations are in third person due to the use of a translator.

## References

Alaii, J. A., W. A. Hawley, M. S. Kolczak, F. O. Ter Kuile, J. E. Gimnig, J. M. Vulule, A. Odhacha, A. J. Oloo, B. L. Nahlen and P. A. Phillips-Howard. 2003. "Factors Affecting Use of Permethrin-Treated Bed Nets during a Randomized Controlled Trial in Western Kenya." *American Journal of Tropical Hygiene and Medicine* 68(4): 137–141.

Baume, C. A. and C. M. Marin. 2007. "Intra-Household Mosquito Net Use in Ethiopia, Ghana, Mali, Nigeria, Senegal, and Zambia: Are Nets Being Used? Who in the Household Uses Them?" *The American Journal of Tropical Medicine and Hygiene* 77(5): 963–971.

Bejon, P., E. Ogada, N. Peshu, and K. Marsh. 2009. "Interactions between Age and ITN Use Determine the Risk of Febrile Malaria in Children." *Public Library of Science ONE* 4(12): 1–5.

Binka, F. N., A. Hodgson, M. Adjuik, and T. Smith. 2002. "Mortality in a Seven-and-a-Half-Year Follow-up of a Trial of Insecticide-Treated Mosquito Nets in Ghana." *Transactions of the Royal Society of Tropical Medicine and Hygiene* 96(6): 597–599.

Chibwana, A. I., D. P. Mathanga, J. Chinkhumba, and C. H. Campbell. 2009. "Sociocultural Predictors of Health-Seeking Behavior for Febrile Under-Five Children in Mwanza-Neno District, Malawi." *Malaria Journal* 8(219).

Chukwuocha, U. M., I. N. S. Dozie, C. O. E. Onwuliri, C. N. Ukaga, B. E. B. Nwoke, B. O. Nwankwo, K. S. Nwoga, et al. 2010. "Perceptions on the Use of Insecticide Treated Nets in Parts of the Imo River Basin, Nigeria: Implications for Preventing Malaria in Pregnancy." *African Journal of Reproductive Health* 14(1): 117–128.

Chuma, J., V. Okungu, and C. Molyneux. 2010. "Barriers to Prompt and Effective Malaria Treatment among the Poorest Populations in Kenya." *Malaria Journal* 9(144).

Chuma, J., V. Okungu, J. Ntwiga, and C. Molyneux. 2010a. "Toward Achieving Abuja Targets: Identifying and Addressing Barriers to Access and Use of Insecticides Treated Nets among the Poorest Populations in Kenya." *BMC Public Health* 10(137).

Chuma, J., V. Okungu, J. Ntwiga, and C. Molyneux. 2010b. "The Economic Costs of Malaria in Four Kenyan Districts: Do Household Costs Differ by Disease Endemnicity?" *Malaria Journal* 9(149).

Cunha, C. B., and B. A. Cunha. 2008. "Brief History of the Clinical Diagnosis of Malaria: From Hippocrates to Osler." *Journal of Vector Borne Disease* 45: 194–199.

Diallo, D. A., S. N. Cousens, N. Cuzin-Ouattara, I. Nebie, E. Ilboudo-Sanogo, and F. Esposito. 2004. "Child Mortality in a West African Population Protected with Insecticide-Treated Curtains for a Period of up to Six Years." *Bulletin of the World Health Organization* 82: 85–91.

Division of Malaria Control (DOMC). 2009. *National Malaria Strategy 2009–2017*. Nairobi, Kenya: Ministry of Public Health and Sanitation.

Division of Malaria Control (DOMC), Ministry of Public Health and Sanitation, Kenya National Bureau of Statistics (KNBS), and ICF Macro. 2011. "2010 Kenya Malaria Indicator Survey." http://www.measuredhs.com/pubs/pdf/MIS7/MIS7.pdf.

Dunn, C. E., A. Le Mare, and C. Makungu. 2010. "Malaria Risk Behaviors, Sociocultural Practices and Rural Livelihoods in Southern Tanzania: Implications for Bed Net Usage." *Social Science and Medicine* 72(2011): 408–417.

Eisele, T .P., J. Keiting, M. Littrell, D. Larsen, and K. Macintyre. 2009. "Assessment of Insecticide-Treated Bednet Use among Children and Pregnant Women across 15 Countries Using Standardized National Surveys." *The American Journal of Tropical Hygiene and Medicine* 80(2): 20–214.

Gatei, W., S. Kariuki, W. Hawley, F. ter Kuile, D. Terlouw, P. Phillips-Howard, B. Nahlen, et al. 2010. "Effects of Transmission Reduction by Insecticide-Treated Bed Nets (ITNs) on Parasite Genetics Population Structure: I. The Genetic Diversity of Plasmodium Falciparum Parasites by Microsatellite Markers in Western Kenya." *Malaria Journal* 9(353): 2–11.

Gill, G. V., and N. J. Beeching. 2004. *Tropical Medicine*. Fifth Edition. Malden: Blackwell Publishing Company.

Githinji, S., S. Herbst, T. Kistemann, and M. A. Noor. 2010. "Mosquito Nets in a Rural Area of Western Kenya: Ownership, Use, and Quality." *Malaria Journal* 9(250).

Health Department. 2010. "Beyond Prevention: Home Management of Malaria in Kenya." http://www.ifrc.org/Global/Publications/Health/Beyond_Prevention_HMM%20Malaria-EN.pdf.

Jambou, R., J. Le Bras, and M. Randrianarivelojosia. 2010. "Pitfalls in New Artemisinin-Containing Antimalarial Drug Development." *Trends in Parasitology* 27(2): 82–90.

Kitzinger, J. 1995. "Qualitative Research: Introducing Focus Groups." *British Medical Journal* 311: 299–302.

Leach, M. and I. Scoones. 2008. "Health Dynamics, Innovation, and the Slow Race to Make Technology Work for the Poor." *Global Forum Update on Research for Health*, 5: 124–127. Pro-Brook Publishing Limited.

Lindblade K. A., T. P. Eisele, J. E. Gimnig, J. A. Alaii, F. Odhiambo, F. O. Kuile ter, and L. Slutsker. 2004. "Sustainability of Reductions in Malaria Transmission and Infant Mortality in Western Kenya with Use of Insecticide-Treated Bednets: Four to Six Years of Follow-up." *Jama* 291: 2571–2580.

Minawaka, N., G. O. Dida, G. O. Sonye, K. Futami, and S. Kaneko. 2008. "Unforeseen Misuses of Bed Nets in Fishing Villages along Lake Victoria." *Malaria Journal* 7(165).

Mungai, M. 2009. "Disease Prevention and Control Program." *Annual Report and Financial Statements 2009*. Kenya Red Cross Society.

Noor, M. A., V. C. Kirui, S. J. Brooker, and R. W. Snow. 2009. "The Use of Insecticide-Treated Nets by Age: Implications for Universal Coverage in Africa." *BMC Infectious Diseases* 9(369).

Ntozi, J. P. M. 1997. "Widowhood, Remarriage, and Migration during the HIV/AIDS Epidemic in Uganda." *Health Transition Review* 7: 125–144.

Patouillard, E., K. G. Hanson, and A. C. Goodman. 2010. "Retail Sector Distribution Chains for Malaria Treatment in the Developing World: A Review of the Literature." *Malaria Journal* 9(50).

Roll Back Malaria Partnership (RBM). 2006. "A Guide to Gender and Malaria Resources." http://www.rbm.who.int/globaladvocacy/docs/gm_guide-en.pdf.

Roll Back Malaria Partnership (RBM). 2008. "The Global Malaria Action Plan." http://www.rbm.who.int/gmap/toc.pdf.

Ruebush, T. K., M. K. Kern, C. C. Campbell, and A. J. Oloo. 1995. Self-Treatment of Malaria in a Rural Area of Western Kenya." *Bulletin of the World Health Organization* 73(2): 229–236.

Tsuang, A., J. Lines, and K. Hanson. 2010. "Which Family Members Use the Best Nets? An Analysis of the Condition of Mosquito Nets and Their Distribution within Households in Tanzania." *Malaria Journal* 9(211).

United Nations (UN). 2011. "Millennium Development Goals Report 2011." http://www.un.org/millenniumgoals/pdf/%282011_E%29%20MDG%20Report%202011_Book%20LR.pdf\.

United Nations (UN). 2012. "Millennium Development Goals Report 2012." http://www.un.org/millenniumgoals/pdf/MDG%20Report%202012.pdf.

United States Ages for International Development (USAID). 2013. "Quicker Malaria Test and Better Policy = Better Healthcare for Kenyans." http://kenya.usaid.gov/success-story/1461.

Webster, J., J. Hill, J. Lines, and K. Hanson. 2005. "Delivery Systems for Insecticide Treated and Untreated Mosquito Nets in Africa: Categorization and Outcomes Achieved." *Health Policy Planning* 22(5): 277–293.

Whitty, C. J. M., C. Chandler, E. Ansah, T. Leslie, and S. G. Staedke. 2008. "Deployment of ACT Antimalarials for Treatment of Malaria: Challenges and Opportunities." *Malaria Journal* 7(1).

Winskill, P., M. Rowland, G. Mtove, R. C. Malima, and M. J. Kirby. 2011. "Malaria Risk Factors in Northeast Tanzania." *Malaria Journal* 10(98): 1–7.

World Health Organization (WHO). 2013. *World Malaria Report 2013.* Geneva: The World Health Organization. http://www.who.int/malaria/publications/ world_malaria_report_2013/report/en/.

# PART II

*Redesigning Standards and Making Public Health Trade-Offs*

# CHAPTER 6

# The Translation of Nutri-Epigenetics into Public Health Policy: The Case of Folic Acid Supplementation

*Maria M.C. Verhagen, Angela Brand, and Elena Ambrosino*

## Introduction

The relationship between folate deficiency and the occurrence of Neural Tube Defects (NTDs) was introduced as early as 1965 (Hibbard, Hibbard, and Jeffcoate 1965). Evidence from scientific studies has since conclusively demonstrated that folic acid (synthetic form of folate) supplementation can prevent the occurrence of NTDs (European Food Safety Authority (EFSA) 2009); its prevalence varies across the European Union (EU) and is currently reported to range from 0.4 to 2.0 per 1000 live births (EFSA 2009). This has led many countries to recommend that women who plan to become pregnant should begin supplementing their diet with folic acid. The issue is that the target group of women who might become pregnant need to consume folic acid in the four weeks prior to conception and eight weeks after conception. This voluntary intervention is difficult to implement effectively because a significant proportion of all pregnancies are unplanned or mistimed. This challenge, along with the discovery in the early 1990s that folic acid protects against cardiovascular diseases later in life (Boushey et al. 1995), led public health prevention policies to develop in two ways (Cornel et al, de Smit, and de Jong-van den Berg 2005): America and some developing countries implemented mandatory fortification of staple foods such as flour (Oakley 2002; Oakley and Johnston 2004), and voluntary fortification of food with folic acid began, and is ongoing, in most European countries. Currently, no EU country has implemented mandatory fortification (EFSA 2009).

Despite mandatory fortification programs leading to increased folic acid intake, research has shown that these programs do not reach all women of reproductive age adequately (Berner, Clydesdale, and Douglass 2001; Imhoff-Kunsch et al. 2007). Furthermore, US studies have shown a dramatic increase in measurements of folate in blood at a population level, postfortification (Ulrich and Potter 2006), raising concerns that fortification exceeds the original daily intake target by as much as two-fold (Choumenkovitch et al. 2002; Shane 2003; Quinlivan and Gregory 2003). At the same time, an advertising hyperbole of a fortified health foods market seems to have occurred, "where a muddying of waters can occur regarding the ideal between too little or too much of any given nutrient" (Lucock and Yates 2009). This is of concern, taking into account the growing evidence in basic science, in particular in nutri-epigenetics, that has recently suggested a possible association between high intake levels of folic acid and the risk of cancer (EFSA 2009), as well as other harmful effects.

In general, according to the European Food Safety Authority, "intervention studies using folic acid have produced a range of different results including adverse effects; overall they do not support the hypothesis that folic acid supplementation of human populations reduces the chronic disease risk" (2009). Effects of folic acid supplementation were reviewed by Lucock and Yates in 2009, with the primary positive effects described as a lower risk for birth defects (MRC Vitamin Study Group 1991). Moreover, evidence from in vitro, animal, and human studies has shown that folate supplementation can also prevent tumor initiation (Choi and Mason 2002; Kim 2005). However, it also seems to facilitate the progression of precancerous lesions (Ulrich and Potter 2007): "Changing the folate status in humans has been shown to influence DNA methylation, but ... it is not yet established whether alterations in DNA methylation after changes in folate status are harmful in humans, for example, by regulating the expression of oncogenes or tumor-suppressor genes" (Smith, Kim, and Refsum 2008). A recent study by Berner et al. (2010) found that exposure to a high concentration of folic acid enhanced cancer cell growth; concomitantly, an increased methylation of estrogen receptors and tumor suppressor promoters was observed, while a lower concentration of folic acid decreased cell growth. Moreover, abnormal promoter hyper methylation, which is associated with inappropriate gene silencing of tumor suppressor genes, is hypothesized to affect virtually every step in tumor progression (Clark and Melki 2002). In addition, the influence of folate status on DNA methylation in both animals and humans is theorized to be tissue-, site-, and gene-specific (Kim 2005; McCabe and Caudill 2005). Studies reviewed by Smith, Kim, and Refsum (2008) show that it may not be justified to assume that protective effects of high folate in a population applies to all individuals in it. Indeed, one cause for concern is that increasing folate levels may, in some people, increase their risk of cancer. Lucock and Yates stated, "In the wider context of folate, it seems to be a question of what level of intake, what form of the vitamin and in the case of cancer patients, when it is supplemented that are the key" (2009).

The interactions between gene expression and folic acid methylation are complex, and the underlying biochemical mechanisms remain elusive; therefore, a population-based prevention method such as folic acid food fortification could be hazardous. "The highly complex and critical biological importance of folic acid-related molecular nutrition makes it a difficult micronutrient to deploy as a simple intervention at a population level - it has far too many biochemical spheres of influence to predict effects in a generalized way" (Lucock and Yates 2009). Public health decisions that involve trade-offs regarding standardized preventive interventions at a population level are common. Epigenetics-based insights are proposed by some scientists to play a major role in showing that individualized solutions to such trade-offs are necessary. Based on insights that the interactions between gene expression and folic acid methylation seem to vary on a personal level, scientists argue for adapting folic acid supplementation to individual genotypes (Van den Veyver 2002) and individual epigenetic DNA methylation profiles. Besides generally healthy food intake, this may be accomplished in the future by using epigenetics-based technology that will allow scientists to adapt to individualized epigenetic contexts.

The public health trade-off consists of whether or not to wait to translate the latest epigenetic evidence into an "evidence-based population level prevention" approach based on research outcomes, which could take decades. In addition, there is growing scientific evidence today that increasing folic acid intake: influences gene expression in exposed populations (Smith, Kim, and Refsum 2008); affects individuals and populations differently based on (epi) genetic characteristics, life stage, and food intake (Lucock and Yates 2009); and can prevent the occurrence of NTD's if taken a certain number of weeks around conception (EFSA 2009).

Regulations for mandatory fortification of flour with folic acid are currently in place in 53 countries worldwide (Centers for Disease Control and Prevention 2010). The EU currently prohibits countries from refusing the import of fortified products, unless these products form a specific danger to public health ((EC) No.1925/2006 and the 2004 decree of the European Court). Intakes of folic acid should not exceed the established Upper Limit (UL) of 1mg/day (EFSA 2009). However, these guidelines can be questioned for a number of reasons: the fortification standards of the products themselves are very unregulated, and the actual maximum levels for the addition of folic acid to foods have not yet been set in European regulations, with data showing that levels vary widely (e.g., the highest levels added to bread spreads up to 1000 µg/100g) (EFSA 2009); European citizens are uninformed about the possible side effects; there is evidence from the UK indicating that some people exceed the UL for folic acid due to voluntary food fortification alone (EFSA 2009), especially children and the elderly; the UL itself is based on limited supporting evidence (EFSA 2009); and scientific evidence already points to possible harmful effects of high folate for some individuals (Smith, Kim, and Refsum 2008).

Currently, individuals cannot be treated with folic acid supplementation based on their personal "susceptibility." According to some scientists, this is a problem considering the fact that it could be harmful for some, based on growing scientific evidence, in particular from nutri-epigenetics. A controversy is created between an existing standard of standardized folic acid supplementation and new scientific nutri-epigenetic insights (folate status influencing DNA methylation) that promise to personalize folic acid supplementation locally. The following chapter investigates how this controversy surrounding the knowledge concerning standardized folic acid supplementation and its potential personally (locally) varied effects gets translated into public health policy. Stakeholders, factors, and relations influence this translation; this chapter examines how these factors do, should, and could impact this interpretation.

A literature review has been performed to investigate the process of translation of nutri-epigenetics in European public health policy regarding the intervention of folic acid supplementation and to identify the relevant actors influencing this process. Through 27 semi-structured interviews (30 to 60 minutes in length), information was collected from key stakeholders about their familiarity of the process of translation of nutri-epigenetics in the current European public health policy regarding the intervention of folic acid supplementation; views on the existing actors and factors influencing the process of translation of nutri-epigenetics in European public health policy; and the challenges in the process of translation of nutri-epigenetics in current European public health policy. The interviewed key stakeholders were selected due to their broad knowledge on, as well as their central roles in, the translation of innovative (nutri-epigenetic) knowledge concerning the public health intervention of folic acid supplementation in public health policy in the Netherlands and Austria. The interviews consisted of open-ended questions to ensure that the interviewees' views were comprehensively explored. The collected data was qualitatively analyzed to identify emerging themes and issues. Overall, we identified four categories of stakeholders in the field: public health risk managers, public health risk assessors, nutritional and (bio)medical experts/risk (technology) assessors, and industrial and ethical institutions. The public health risk managers are responsible for decision making and translation of nutri-epigenetics concerning public health activities. Interviewees worked at different national authorities (e.g., Ministry of Health, Welfare and Sports in the Netherlands, and Ministry of Health in Austria). These risk management actors make their policy decisions partly based on the independent advice of their relevant advisory public health risk assessors. Actors in this second category of interviewees included professionals from the Dutch Health Council in the Netherlands, the Austrian Society of Nutrition, the Commission of the Austrian Food Nutrition Action Plan, the Ludwig Boltzmann Institute for Health Technology Assessment in Austria, and the 2009 EFSA report as the keystone of European Union (EU) risk assessment regarding food and feed safety. The third category, nutritional and (bio)medical experts, consisted of scientific experts from the field of nutritional and (bio)medical science who are consulted by the public health risk assessor institutions. For the interviews, experts from the following fields were selected: nutrigenomics; (human) nutritional science; health nutrition and environmental protection; systems biology; genetics and cell biology; experts in cancer and folic acid research, community genetics, and public health genomics; medicine and society; gynecology; pediatrics; agencies for health, food, and consumer product safety; institutes for health technology assessment; risk assessment/modeling/monitoring; clinical epidemiology; center for society and genomics; and editors of journals for nutritional medicine.

The chapter shows that the latest insights in nutri-epigenetics call for a more thorough understanding of individualized (folate) needs. Yet, these promises clash with established public health notions of risk/benefit, cost-effectiveness, and standardization. The analysis shows how difficult it is to introduce a "personalized standard" into public health, a sociotechnical setting that is traditionally operates using "universal standards" for the whole population.

## Translating Nutri-Epigenetics Insights into Folic Acid Supplementation Policy

How are nutri-epigenetic insights being translated into public health policy with regard to folic acid supplementation? Most policy documents and stakeholders did not mention or discuss nutri-epigenetics in relation to folic acid supplementation policy. Yet, most stakeholders had heard about the main epigenetics mechanism, namely methylation of DNA, and could also link this mechanism with the working mechanisms of folic acid. Further, they mentioned many more (bio)medical working mechanisms that could interact and contribute to (harmful) effects of high folic acid levels. Many stakeholders also mentioned their concerns regarding a rising level of folic acid in food products and supplements and, therefore, unmetabolized folic acid in the population's blood levels. Hence, most stakeholders discussed a fear of cancer and unpredictable intergenerational effects. These effects, currently discussed and taken into account during policy decisions, are partly associated with nutri-epigenetic mechanisms; one could argue that translation of nutri-epigenetics knowledge with regard to folic acid is, at least partially, taking place at this moment.

Stakeholders' discussions about the scientific term "nutri-epigenetics" in relation to folic acid supplementation policy and/or the importance of beginning research in this specific field seemed to be related to the way in which people defined nutri-epigenetics. Stakeholders' definition of nutri-epigenetics also affected how they viewed this field: some saw it as being a "new" field, and others viewed it as part of the overall (bio)medical working mechanisms of folic acid in the human body.

> *I do not exactly know what is different about epigenetics ... I do think it creates new insights ... new factors that play a role [in gene-environment interaction] ... yet I still have to see with my own eyes that we can actually do something with it [new insights] ... the influence of those different factors is so small ... we work with stratified [in contrast to individualized] medicine for a long time already, because we can never proof anything on an individual level, you never know in which phase and in which cells [happens what] ... you never have scientific proof for an individual ... in practice we always measure in big [substantial] groups of people."*
> *(Interview, expert in clinical epidemiology, EMC Rotterdam, 08-15-2011).*

Most stakeholders did not know the details of the working mechanisms of folic acid, as it was not part of their job. However, they did care about the effects these mechanisms have on health; they especially cared about the "beneficial," "toxic," or "harmful" levels of the substance.

> *There you touch upon an interesting point, because toxicology and also nutrition works with everything we can see with our own eyes ... I think it is important to recognize that when we are talking about the relation of FA [folic acid] and Neural Tube Defects or cancer, that when we can clearly show an existing effect, for good or for bad, then we can do something with it. But the fact that there are all sorts of [biomedical] mechanisms behind this final effect, does not interest me: I don't care. (Interview, public health expert in nutrition and health, RIVM, Bilthoven, 08-17-2011)*

The complex working mechanisms of folic acid in the human body tend to become more complex when one goes into the field of nutri-epigenetics The level of expertise on the subject seems to create a difference in stakeholders' understanding of the "risks" of a rising population's blood level of unmetabolized folic acid. To illustrate, most nutritional and biomedical scientists generally feared the harmful effects of high folic acid intake and the intergenerational effects in the future.

*Too much FA does not automatically mean that more methylation will take place ... but, you should not let it happen of course, because then it is too late. I myself find it a very dangerous action [population wide FA food fortification]; irreversible [epigenetic] set points in the human development phases do exist and if you modify those mechanisms, you are busy with humongous human experiments without any permission. (Interview, expert in nutritional sciences, Univ. Wageningen, 08-16-2011)*

In contrast, many stakeholders involved in public health risk-assessment or risk-management seemed more focused on the balance of both beneficial (decrease of NTDs) and harmful (risk of cancer) effects for the population as a whole over a relatively short period of time, as opposed to "possible" harmful intergenerational effects.

*If it is based on evidence that FA supplementation leads to a significant decrease in NTDs and it is a vitamin and does not have many harmful effects, I would argue: let's start stimulating a little more FA intake, especially stimulate the subgroups who need it most, and I am not immediately very scared for possible biochemical harmful effects on health." (Interview, expert in community genetics and public health genomics, VU Amsterdam, 08-03-2011)*

The following paragraph further discusses the roles of the actors identified in this study and illustrates how their variety of goals, decisions, conflicts, and political and economic contexts influenced and/or hindered the process of the translation of nutri-epigenetics into public health policy with regard to folic acid supplementation.

### The Innovation Network of Nutri-Epigenetics and its Diversity of Goals and Conflicts

Many nutritional and (bio)medical experts argued for a preventive approach focused on healthy (biological) food production and consumption in general; this approach leads to a personalized balance of the metabolism and methylation-pattern in the long term.

*It is always easier to think of a short-term solution [fortification] than to thoroughly look into: what is wrong in our food chain in general? From production of food till the consumption of food. Because these are structural solutions that demand long-term [policy] changes. (Interview, expert in society sciences, Univ. Wageningen, 08-11-2011)*

These experts worried that many food products tend to get modified (fortified with folic acid) based on insufficient knowledge, leading to unnatural eating habits; insufficient knowledge could lead to diseases now and in future generations, partly due to epigenetic mechanisms.

*I find it extremely unbiological and not natural to select one B vitamin {FA} from all the other vitamins, while in the metabolism and DNA- methylation process it is not only this vitamin that plays a role it is also B2, B12, B6 ... and if we would fortify our bread with only FA one could hypothesize that one would disrupt an important naturally balanced equilibrium ... a tomato is a 'tomato' for a reason ... it also works in a different way, we accumulate this [unmetabolized] substance [FA] in our body [blood] and the question is if this does [not] have any harmful effects in the future? Actually we do not know enough about it to make any informed decisions about FA-fortification. (Interview, expert in nutrition and cancer, Univ. Wageningen, 08-23-2011)*

In contrast, some risk assessors referred to (biomolecular) genetic evidence derived from research and past experience to make a risk/benefit assessment for the population as a whole, while emphasizing the importance of monitoring its effects on health.

*Some say that: if by now, we reached an equilibrium that is adapted to our {former} nutrition habits and suddenly one starts to fortify food, then one can disrupt this equilibrium ... well, do I think this could be harmful? ... From what I know about molecular genetics I would say that such an equilibrium exists so that we can keep our population in a fairly good shape on the long term [during many generations] and to have the possibility to react to changing nutritional circumstances. And those circumstances changed a lot in the last 30 years [about one generation], and the intake of folate is decreasing, so I would rather argue that we could better start to fortify with a little FA ... However, I think it is very important to monitor its effects on health. (Interview, expert in community genetics and public health genomics, VU Amsterdam, 08-03-2011)*

Risk managers need to make practical risk/benefit choices for public health policy during their time of service, which is often over a relatively short period of time.

*Originally I am an expert in the field of nutrition science ... [yet] I became quite skeptical about the fact that people will [ever] start to eat more healthy, it is very hard to make this happen ... and the Ministry of Health Welfare and Sports also has the responsibility to decrease deficiencies [for example, because of low FA vegetable intake] when possible. (Interview, public health authority expert in nutrition and food safety, Min. HWS Den Haag, 08-08-2011)*

Therefore, risk-managers seemed primarily worried about how to most effectively divide their money into health research, prevention, and care interventions. This division of funds leads to the most beneficial and least expensive population health intervention at present and in the future. As such, it was very difficult to base public health folic acid supplementation policy decisions on "evidence" that, in the case of a dynamic (epigenetics) process, is constantly subject to change.

*You can estimate the benefits and the potential risks of a certain policy, weigh them and then choose: what intervention has priority? And this has to be funded by evidence-based data, this is always a problem, cause good data needs time, there is always a delay, I do think that one should not handle too fast, ten years ago people found it ridiculous that we did not fortify folic acid more in the Netherlands because it was supposed to be 'good' for prevention of heart disease." (Interview, expert from the nutrition center, Den Haag, 08-19-2011)*

This forced public health authorities to cope with difficult decisions: Should a preventive intervention of folic acid supplementation be focused on an individual, on a subgroup, or at the population level? Will future research bring any further solutions and benefits? How do authorities act wisely at this very moment?

*So the people responsible for public health are in a terrible dilemma. Namely, it is complicated [to make the most "right" policy decision with regard to FA supplementation, taking into account the newest epigenetic evidence/harmful effects on health]; thus, we need more research ... yet the more knowledge we get, the more we tend to think in population subgroups and then the question arises: Should we bother the [individual] consumers with this kind of complicated things and decisions, or shall I simply make the choices for them? ... I think that public health authority is acting wisely by saying: we take it slowly, we do not fortify with a high amount, we search for subgroups with clear pathology and need for FA. (Interview, expert in systems biology, TNO Zeist, 08-01-2011)*

Further, it was often cited that one should not underestimate the economic constraints that seem inextricably linked with decisions regarding public health food fortification.

*I understood that this decision [not to mandatorily fortify] is not really made based on scientific evidence, but because of the fact that the costs for monitoring of the entire Dutch population, to check folic acid levels, seemed too expensive to the government. (Interview, expert in nutrition and cancer, Univ. Wageningen, 08-23-2011)*

Additionally, public health decision making in general was intricately connected to its global economic situation, political contexts, and legislation.

*The Health Council is at that moment concerned with the "piece" of folic acid only, and concerned with which population subgroups have a need for this vitamin, however, this is not the whole story ... the story is much more complicated: it is about decisions concerning the food chain in general ... why is it possible that the decision is made to fortify foods with nutrients that do not belong in those products in the first place? This decision making is more complex than a scientific report ... politics is economy, and health for sure is. (Interview, expert in society sciences, Univ. Wageningen, 11-08-2011)*

Important actors in political decision making regarding food fortification are the food and (bio) pharmaceutical industries. Public health managers concerned with folic acid supplementation policy are part of a national and European government; they must take into account many national and European trade regulations.

*Decision making at the Codex [Alimentarius] ... it does not have much to do with science, it has to do with trade barriers ... to keep the economy running we need the big manufacturers and that we can earn money with them. (Interview, expert in society sciences, Univ. Wageningen, 08-11-2011)*

Many nutritional and biomedical scientists, however, worried about the fact that European trade regulations concerning voluntary fortification of food did not necessarily favor public health.

*Kellogg's did win the case [arrest van het Hof van Justitie EG van 2 December 2004, zaak C-41/02] at the time... and was since then allowed to sell fortified cereals on the [Dutch] market ... Do I think that these fortified cereals are an innovative improvement with regard to the Dutch diet? No, I do not think so; however, there are insufficient legal arguments to resist these products on the European and national markets. (Interview, expert from nutrition center, Den Haag, 08-19-2011)*

Nevertheless, there seems to be an upward trend in the growth of the fortified health food market and products that could theoretically be used for "personalized nutrition" purposes. Different types of breakfast cereals for epigenetically different family members, partly based on new individualized epigenetic insights, is one example. Likewise, the (bio) technological industry is interested in creating epigenetics-based personalized tests and treatments. At this moment, however, it seems too early to actually utilize and sell these products.

*A few thousand personal genomes are already on the [inter]net, and the challenge will be to do something useful with it, old school research ... your genome is not your destiny, keep that in mind ... epigenetics as the rising star with regard to the interactions with your environment ... these interactions will determine your chances and risks in the future etc., yet you can direct a part of the process, because some methylation patterns are definite yet others are greatly variable, people tend to get more aware of this notion I would say ... the cancer genomics center is working toward a more personalized way of therapy: tumor diagnostics, developments in medicine, and targeted drugs based on molecular points of action. (Interview, public health authority expert from the Center for Nutrition and Health, RIVM Zeist, 08-10-2011)*

## Aligned and Conflicting Goals

One point upon which many nutritional, (bio)medical, and public health scientists agreed was the fact that people have individual (epi)genetic-based needs. Therefore, they opted against a population-wide folic acid fortification of food, primarily based on the assumption that it can harm some or many people in the future. On the other hand, many experts from these fields disagreed that personalized fortified foods were necessary to adequately handle individualized needs in the population.

Many stakeholders established that a natural, balanced diet and lifestyle were the best ways to prevent disease now both and in the future.

*In the Netherlands we start from the principle: if everyone eats varied and balanced, no one is assumed to have a deficit of vitamins [folate], except for a few exceptional population sub groups, like women wishing to get pregnant, therefore we supplement them with FA. (Interview, public health authority expert in nutrition and food safety, Min. HWS Den Haag, 08-08-2011)*

The same stakeholders argued that fortified products could help some, possibly many, individuals in a relatively short time frame and would definitely decrease the medical, psychological, and financial "burden" of NTDs. Many stakeholders also recognized that fortified products can lead to high levels of unmetabolized folic acid in blood and unnatural eating habits in the population, leading to disease and harmful and unpredicted intergenerational effects. Nevertheless, most scientists require money to be able to execute research; these funds partially come from the food/supplement/(bio)technological industry. The respondents of this study acknowledged their potentially vested interests in funding research that supported their claims. The results of this research could then be used by public health risk-managers and risk-assessors to formulate their "informed" decisions.

*Nutrition science is coming from the pediatrics and is supported by Nestle and so on, it is a conflict of interest, I think nutrition is in the hand of a few nutritional companies ... and they give you the information, what you want. (Interview, expert in gynecology, Vienna, 07-22-2011)*

The industry that invests in research for "individualized" (food/supplement/ (bio)technological) products that could be helpful in a "personalized medicine" approach also aims to develop products intended to be sold to as many consumers as possible, therefore targeting the population as a whole.

*The European regulation on voluntary fortification ... became more and more liberal, in former times (five till ten years ago) one could only add something to a food product if proved that the addition would have a positive effect on public health, and one could almost never prove that ... however, food companies heavily protested against this regulation and proposed to turn it around [addition to food products as long as harmful effects cannot be strongly proven], because that is more convenient for the food industry because they can sell 'enriched' products [that sell better], like candy with vitamin C. (Interview, public health authority expert from the Dutch Health Council, Den Haag, 06-07-2011)*

Therefore, the industry is pushing trade laws and regulations toward "a right to fortify," as long as harmful effects cannot be quantified. It does appear quite conflicting that this "proof" must be derived from research that is, to a large extent, funded by the industry itself.

*On the base of this regulation [arrest van het Hof van Justitie EG van December 2, 2004, case C-41/02] the trade of fortified foodstuffs can only be prohibited provided that the foodstuffs are harmful to health. More radical regulations lead to barriers for trade prohibited by EG-regulations. I will work actively for the possibility for EU-member states to prohibit fortification of foodstuffs without a nutritional necessity. (Netherlands Ministry of Health, Welfare, and Sport 2010)*

Overall, in the complex network concerning the translation of nutri-epigenetics into public health policy with regard to folic acid supplementation, there is a variety of aligned and conflicting goals that are strongly influenced by political and economic factors. The main points of conflict included disagreements that: a natural, balanced diet and lifestyle are the best ways to prevent disease now and in the future; individual benefits and risks of fortified products in the short and long term appear to exist; and industry is pushing trade laws and regulations toward "a right to fortify," as long as harmful effects cannot be quantified. Additionally, it is not always clear which evidence best establishes the risks and benefits of food supplementation. There are also ongoing discussions on whether investment in natural healthy folic acid intake is the best policy.

Policymakers seemed quite skeptical about the possibility of changing the general population's healthy food consumption habits, especially in a short timeframe. Many (bio)medical experts expressed their concerns about food products being modified, which may lead to unnatural eating habits and possible harmful effects in the future. Other experts in the (community) genetics field inferred that human genetics are naturally adapting to changing nutritional circumstances. Concerns were raised, predominantly by risk managers and risk assessors, that most people today have already acquired unnatural eating habits; in the interest of "public health," it could be more cost-effective to stimulate extra folic acid supplementation.

All stakeholders appeared to agree to base any decisions on folic acid food supplementation on scientific evidence, yet they have to deal with the fact that this evidence is constantly changing with fluctuations in the population. This runs counter to the public health attempt to standardize guidelines for an entire population that are valid for at least several years. Most stakeholders interviewed agreed to make decisions on food supplementation slowly, to closely monitor the effects, and to focus on the subgroups or even individual people who need extra intake. However, policymakers also emphasized they have to deal with economic considerations, market dynamics, and alignment with national and international policies and legislation; these issues are demonstrated by authorities being unable to prohibit voluntary food fortification unless harmful effects of the additive(s) can be quantified. These concerns show the challenges of introducing local adaptation of a universal supplementation standard to different population groups or even individuals. It also reveals the clashes between the emerging potential to localize a public health standard and established public health notions of risk/benefit frameworks, cost-effectiveness, and attempts to standardize guidelines for years to come.

## Policy Actions with Regard to Folic Acid
## Supplementation in Europe

In both the Netherlands and Austria, the national health authorities have chosen to focus on a targeted folic acid supplementation policy for women who want to become pregnant; however, both countries cannot prohibit voluntary food fortification. These decisions seem to be based on a conflicting trade-off between liberal trade regulations, which are partly enforced by the European Commission, and local cultural values that do not favor regulated, mandatory fortification. In the Netherlands, for example,

> At this moment we cannot prohibit voluntary food fortification because we opened the European trade frontiers ... On top of that, the Dutch culture and political driving force do not favor regulated mandatory fortification ... in such a situation one looks for other solutions, the European legislation did not set a maximum level of voluntary fortification of foods yet, so as a nation you have to set your own maximum level ... but because Dutch state-politics do not like to see too much administrative burden on the industry, we do not know exactly what products are being fortified within the exemption policy that we enforced ... but for us it is really the question: what about the [young] children, with a low Upper Limit, my first reflex is: do we not cross this level? (Interview, public health authority expert in nutrition and food safety, Min. HWS Den Haag, 08-08-2011)

A Dutch Ministry of Health, Welfare, and Sports-directed study was undertaken to establish the maximum level of folic acid fortification utilizing a calculation model. The model requires numerous parameters on folic acid intake and the associated biological processing by the body. This study also evaluated the risk of national young children exceeding the Upper Limit for folic acid. The following three statements were derived from this study:

> Based on scenario calculations using Dutch National Food Consumption Status [DNFCS]-kids and results from the inventory, currently, there seems to be no risk regarding too high folic acid in young children. Only a small proportion (<5%) of the Dutch children currently have folic acid intakes above the UL. (Verkaik-Kloosterman et al. 2009)

> When the observed trend of increased number of foods being fortified in similar food groups will continue, especially young children (2-3 years) will have a risk of too high folic acid intakes; in that case a chance of harmful health effects cannot be ruled out. (Verkaik-Kloosterman, et al. 2009)

> Based on data from DNFCS-kids, folic acid intake from food supplements is high. Especially among children aged 2-3 years old, intake levels above the UL by food supplements only are observed ... With changing fortification practices, but also changing dietary patterns, the calculation model parameters to calculate

*the maximum fortification levels need to be regularly evaluated and when needed updated.* (Verkaik-Kloosterman, et al. 2009)

The Dutch policy illustrates that the market of products fortified with folic acid is growing without strict monitoring and authorities are unable to prohibit voluntary food fortification; therefore, it is very hard to calculate and control the risks of folic acid intake that is too high, especially in young children. In addition, limited detailed knowledge of the working mechanism of folic acid on the human body at a nutri-epigenetic level means that possible harmful effects of folic acid supplementation cannot be fully evaluated in risk calculation models. Hence, further research would be required to enable the translation of nutri-epigenetics into public health risk calculation models. However, it is unclear whether there will ever be enough evidence to support an intervention that is targeted at the whole population. Some stakeholders suggest to instead focus research and evidence making on clarifying which types of individuals are likely to benefit from folic acid supplementation.

## Discussion and Conclusion

Emerging evidence in nutritional systems biology, in particular in the area of nutri-epigenetics, calls for a paradigm shift in public health regarding standardized, preventive supplementation interventions at a population level. In the case of folic acid fortification, insights from nutri-epigenetics promise to understand individualized folate needs and tailor consumption of products rich in folate (whether fortified or not) in a cost-effective way; this practice will lead to the adjustment of the concept of standardized prevention, which takes a more personalized (localized) approach. This chapter discussed the difficulties in translating such a localized standardization approach with regard to folic acid supplementation in public health policy. In a complex network concerned with the process of the translation of nutri-epigenetics into public health policy, the different perspectives of stakeholders and a variety of aligned and conflicting goals, strongly influenced by political and economic factors, interacted. The most important stakeholders identified were nutritional and (bio)medical experts; public health risk assessors and managers; national and European health authorities; and the food, (bio)pharmaceutical, and (bio) technological industries. The main findings from the interviews were related to the issue of what counts as "right science" when settling the controversy about folic acid fortification and the difficulty to implement this new innovation in a sociotechnic setting that is characterized by a conservative mind set.

It is difficult to assess what is the "right" science when changing established standards, what counts as evidence, and how this evidence is used to settle the controversy. There is disagreement among stakeholders as to whether people have individual epigenetic needs and whether natural, balanced diets and lifestyles are the best ways to prevent disease now and in the future. Most interviewees agreed that important and critical knowledge gaps exist with regard to the working mechanisms of folic acid in the human body and its possible harmful effects. Yet, it is unclear whether there will ever be enough evidence to support an intervention that is targeted at the whole population. Some stakeholders suggest focusing research and evidence making on clarifying which types of individuals are likely to benefit from folic acid supplementation.

It is also challenging to deploy personalized standards in the sociotechnical context of public health that is fueled by standardization efforts with regard to entire populations. EFSA stakeholders, for instance, think in terms of a risk/benefit framework in which cost-effectiveness considerations make the most sense, given the disagreement on individual benefits and risks of fortified products in the short and long term. Public health policy makers operate within a paradigm of population-based thinking and as such they highlighted economic considerations, market dynamics, and alignment with national and international policies and legislation. In such an environment, personalized policies are difficult to introduce.

The implications are that in order to generate and implement a localized standard, such as folic acid supplementation that is based on nutri-epigenetic insights, established institutions that make up the sociotechnical setting of public health policy making (in this case public health authorities, RIVM, EFSA) need to be attuned to this different way of "doing standardization" as well.

## References

Berner, C., E. Aumüller, A. Gnauck, M. Nestelberger, A. Just, and A. G. Haslberger. 2010. "Epigenetic Control of Estrogen Receptor Expression and Tumor Suppressor Genes Is Modulated by Bioactive Food Compounds." *Ann Nutr Metab* 57(3–4): 183–189. doi:10.1159/000321514.

Berner, L. A., F. M. Clydesdale, and J. S. Douglass. 2001. "Fortification Contributed Greatly to Vitamin and Mineral Intakes in the United States, 1989–1991." *The Journal of Nutrition* 131(8): 2177–2183.

Boushey, C. J., S. A. Beresford, G. S. Omenn, and A. G. Motulsky. 1995. "A Quantitative Assessment of Plasma Homocysteine as a Risk Factor for Vascular Disease. Probable Benefits of Increasing Folic Acid Intakes." *The Journal of the American Medical Association* 274(13): 1049–1057.

Centers for Disease Control and Prevention. 2010. "CDC Grand Rounds: Additional Opportunities to Prevent Neural Tube Defects with Folic Acid Fortification." *Morbidity and Mortality Weekly Report* 59(31): 980–984.

Choi, S. W., and J. B. Mason. 2002. "Folate Status: Effects on Pathways of Colorectal Carcinogenesis." *The Journal of Nutrition* 132(8 suppl): 2413–2418.

Choumenkovitch, S. F., J. Selhub, P. W. Wilson, J. I. Rader, I. H. Rosenberg, and P. F. Jacques. 2002. "Folic Acid Intake from Fortification in United States Exceeds Predictions." *The Journal of Nutritio* 132(9): 2792–2798.

Clark, S. J., and J. Melki. 2002. "DNA Methylation and Gene Silencing in Cancer: Which Is the Guilty Party?" *Oncogene* 21(35): 5380–5387.

Cornel, M. C., D. J. de Smit, and L. T. W. de Jong-van den Berg. 2005. "Folic Acid—the Scientific Debate as a Base for Public Health Policy." *Reproductive Toxicology* 20: 411–415. http://www.sciencedirect.com/science/article/pii/S089062380500119X.

European Food Safety Authority (EFSA). 2009. "ESCO Report Prepared by the EFSA Scientific Cooperation Working Group on Analysis of Risks and Benefits of Fortification of Food with Folic Acid." http://www.efsa.europa.eu/en/scdocs/doc/3e.pdf.

Hibbard, B. M., E. D. Hibbard, and T. N. Jeffcoate. 1965. "Folic Acid and Reproduction." *Acta Obstetrica et Gynecologica Scandinavica* 44(3): 375–400.

Imhoff-Kunsch, B., R. Flores, O. Dary, and R. Martorell. 2007. "Wheat Flour Fortification Is Unlikely to Benefit the Neediest in Guatemala." *The Journal of Nutrition* 137: 1017–1022.

Kim, Y. I. 2005. "Nutritional Epigenetics: Impact of Folate Deficiency on DNA Methylation and Colon Cancer Susceptibility." *The Journal of Nutrition* 135(11): 2703–2709.

Lucock, M., and Z. Yates. 2009. "Folic Acid Fortification: A Double-Edged Sword." *Current Opinion in Clinical Nutrition and Metabolic Care* 12(6): 555–564. doi:10.1097/MCO.0b013e32833192bc.

McCabe, D. C., and M. A. Caudill. 2005. "DNA Methylation, Genomic Silencing, and Links to Nutrition and Cancer." *Nutrition Reviews* 63(6 Pt 1): 183–195.

MRC Vitamin Study Group. 1991. "Prevention of Neural Tube Defects: Results of the Medical Research Council Vitamin Study." *Lancet* 338(8760): 131–137.

Netherlands Ministry of Health, Welfare, and Sport. 2010. "Voedingsbeleid; Brief Minister over Adviezen van de Gezondheidsraad over Microvoedingsstoffen." Accessed May 12. 2011. http://m.europa-nu.nl/9353000/1/j9vviaekvp0oeyh/vibs9q1dt6xr.

Oakley, G. P. 2002. "Inertia on Folic Acid Fortification: Public Health Malpractice." *Teratology* 66(1): 44–54.

Oakley, G. P., and R. B. Johnston. 2004. "Balancing Benefits and Harms in Public Health Prevention Programs Mandated by Governments." *The British Medical Journal* 329(7456): 41–43.

Quinlivan, E. P., and J. F. Gregory. 2003. "Effect of Food Fortification on Folic Acid Intake in the United States." *The American Journal of Clinical Nutrition* 77(1): 221–225.

Shane, B. 2003. "Folate Fortification: Enough Already?" *The American Journal of Clinical Nutrition* 77: 8–9.

Smith, A. D., Y. I. Kim, and H. Refsum. 2008. "Is Folic Acid Good for Everyone?" *Am J Clin Nutr* 87(3): 517–533. http://www.ajcn.org/content/87/3/517.long.

Ulrich, C. M., and J. D. Potter. 2006. "Folate Supplementation: Too Much of a Good Thing?" *Cancer Epidemiol Biomarkers Prev* 15(2): 189. doi: 10.1158/1055-9965. EPI-06-0054.

Ulrich, C. M., and J. D. Potter. 2007. "Folate and Cancer—Timing Is Everything." *The Journal of the American Medical Association* 297(21): 2408–2409.

Van den Veyver, I. B. 2002. "Genetic Effects of Methylation Diets." *Annu. Rev. Nutr* 22: 255–282. doi: 10.1146/ annurev.nutr.22.010402.102932.

Verkaik-Kloosterman, J., Tijhuis, M. J., Beukers, M., & Buurman-Rethans, E. J. M. (2009). Evaluation of the Dutch legislation on food fortification with folic acid and vitamin D; focus on young children. RIVM Letter report 350090006/2009. Retrieved from http://www.rivm.nl/bibliotheek/rapporten/350090006.pdf.

# CHAPTER 7

# Social Confounders of Vaccine Response

*MeiLee Ling, Angela Brand, and Elena Ambrosino*

## Introduction

Among public health initiatives, vaccination has played an important role in improving human health. Every year, it is estimated that vaccines prevent about six million deaths worldwide (Ehreth 2003). In the US, vaccines such as MMR (Measles, Mumps, and Rubella), DTP (Diphtheria, Tetanus, and Pertussis), and Hib (Haemophilus influenza type B vaccine) have been responsible for a 99 percent decrease in incidents of those diseases. This decrease has resulted in a similar decline in mortality and disease sequels from 1990–1998 (Centers for Disease Control and Prevention (CDC) 1999). However, the concept of "one size fits all" in the vaccination field has been used in populations with very different characteristics (Kennedy and Poland 2011; Poland, Ovsyannikova, and Jacobson 2008). The assumption is made that most people will respond and develop antibodies in the same way; therefore, the same vaccine doses are administered equally to different groups of people. This practice overlooks individual and subpopulation differences in immune responses, such as poor vaccine response due to inadequate uptake of the antigen or failure of the immune system to produce the appropriate antibody response. Such differences are partly due to the complex interaction between biological, genetic, social, and environmental factors in immune response. Although the final goal of vaccine research is ideally to develop a universal vaccine that provides standardized protection to various populations across the world, inconsistent outcomes of vaccinations have been observed across populations with different features. A study conducted in the UK and Malawi, regarding the Bacillus Calmette-Guérin (BCG) vaccination against tuberculosis (TB), has shown that the "one size fits all" approach is unlikely to be the best option. The same study reported that since the BCG vaccination was introduced in the UK adolescent population, the protection against pulmonary TB has increased from 50 to 80 percent. However, no protection was observed when the same BCG vaccine was used in Malawian adolescents (Black et al. 2002). It was suggested that the variability in BCG efficacy is the effect of the populations' differential sensitization due to environmental mycobacteria exposure. Thus the universally standardized vaccine is not effective in every locality.

Throughout the history of vaccinology, the development of vaccines has relied on trial-and-error experimental processes that may influence the pathway of vaccine response; the early events of immunity development remain unknown. The task of vaccine development focuses on reducing a pathogenic microbe to its smallest immunogenic component. It then relies on the immune system to build a protective immune response against it. Later in the process, the same immune system will destruct the exogenous microbic particles. The success of vaccine development is dependent on the human responses at the end of clinical trials. If the experimental vaccines do not induce the expected protective and sustained immune response, alternative antigen candidates are selected or new adjuvant delivery systems are formulated and developed; the developers then begin again with trial and error. An additional challenge in the development process is that the human immune system is highly complex and the environmental factors that change the intrinsic capacity of the recipient's response to a vaccine remain mostly unknown (Bernstein, Pulendran, and Rappuoli 2011).

Once the vaccine has been introduced into a population, every characteristic of the population plays a role in the human biological pathway of vaccine response. Such characteristics could play an important role in biological processes in the global setting. Vaccines are critically important to human health, especially in low- and middle-income countries (LMICs); further research in the field of vaccine efficacy and socioenvironmental factors is required to improve the process of developing vaccines for the benefit of different populations.

For a long time, immunological research has focused on the interplay among biological, behavioral, and social factors in health and disease. More recently, the influence of social factors on the immune system has been investigated (Azad et al. 2012; Cohen and Herbert 1996; Dowd et al. 2008; Ponton et al. 2011). However, similar research on the relationship between the social environment and responses to vaccines is currently lacking. At present, there are no extensively validated approaches for studying the effect of social variation on vaccine response. However, the literature has shown that in many cases, variation in vaccine-induced immune response is also associated with social variation. To gain a better understanding of the human response to vaccines, we should take other variables (social confounders) into consideration that could influence the immune response, such as socioeconomic status, nutrition, and vaccination history. Ignoring differences of socioenvironmental factors among populations can affect the vaccine efficacy and may lead to failure in disease prevention programs. One important new approach to address the social confounders influencing the human immune response to vaccines is "environ-vaccionology," which examines the relationship between immune response to vaccines, environmental and social determinants, and how these are applied to vaccine research and development. Environ-vaccinology involves global population-based studies and includes the consideration of differences in sociocultural and biological-physical factors across groups and their relationship to the variation in vaccine responses. A system biology approach is taken in order to gain a deeper understanding of the complex interactions between the different components of this process. Environ-vaccinology aims to incorporate multiple levels of knowledge on social and environmental variation, as well as pathways of interaction with host biological factors. The goal of environ-vaccinology is to predict population responses to vaccines and to determine the most efficient vaccine types, as well as the most suitable course of vaccination. Environ-vaccinology introduces social science elements into the vaccine discovery process. The collaboration between multidisciplinary teams of laboratory, computational, and clinical scientists with social scientists is crucial to understand sociocultural factors related to a certain community and to map the relationship between environmental factors and immune responses. For example, to better understand the biological effects of certain environmental factors, the identification of host-specific factors is important. Therefore, population-based epidemiologic studies could be helpful to explore possible social determinants that influence human immunological pathways such as measuring associations between environmental factors (e.g., diet, socioeconomic status, disease history, vaccination history) and vaccine responses.

The emerging scientific field promises a research approach that introduces localized elements (social confounders) into a technical standard (vaccine). In this chapter, we will review these innovations in vaccine development to explore the state-of-the-art in vaccine development regarding social confounders on vaccine response as well as its possible application into global public health policy.

A qualitative research approach, combining a literature review with open-ended interviews, was used to explore perspectives by public health experts on environ-vaccinology and the opportunities and challenges in applying this innovation in the field of vaccinology. Qualitative research was particularly valuable in this case, as the topic of environ-vaccinology has been minimally investigated and its key dimensions remain largely unknown. The literature review identified publications specific to social factors and vaccine response. The following databases were searched: Web of Science, PubMed/Medline, and Google Scholar. Search terms included social factors and vaccine response, vaccine failure and environmental factors, vaccination history and vaccine response, immunity background and social factors, nutrition and vaccine response, as well as public health genomics translational framework. All search results were scanned for relevance and methodological quality, and information from relevant publications was used to develop a theory of social confounders on vaccine response. Next, semi-structured interviews were completed with six public health officials and vaccine experts from the European Center for Disease Control and Prevention (ECDC). These interviewees were selected due to their broad knowledge of vaccinology and the vaccine industry, as well as their central roles in the field of public health. Additional individuals in the ECDC identified during the interview process were also considered for recruitment. The interviews consisted of open-ended questions to ensure that the interviewees' views were comprehensively explored.

The collected data was qualitatively analyzed to identify emerging themes and issues. Questions that were asked during the interviews included: What are the innovations in vaccine development regarding social confounders and how are those changing the protocol of vaccine development? What practices, national policies, and other contextual factors can contribute to the success of environ-vaccinology applications? What can public health officers and decision makers learn from these innovations in vaccine, and what issues concern them? Since this involved exploratory research into innovations in vaccine development, there is some limitation in the study related to the setting constraints; only a sample of six interviews with members of the ECDC have been performed. It may have been useful to also gain perspectives of experts from other nongovernmental agencies and educational institutions as well as from members of the broader community. Furthermore, the working environment and characteristics of this ECDC agency may draw some biases from the experts. We will first discuss the results from the literature review regarding social confounders on vaccine response. The second section includes the analysis from the interviews with public health and vaccine experts.

## Influence of Social Confounders on Vaccine Response: A Literature Review

A range of studies in the field of vaccine study has been identified that discusses possible social confounders on vaccine response ranging from socioenvironmental factors—such as disease history, nutritional status, socioeconomic status, and vaccination history—to sex and gender of a population. In vaccine immunogenicity studies, the research usually takes into consideration host heterogeneity to avoid bias in the vaccine efficacy results (Flanagan et al. 2010; Flanagan,et al. 2013). The study from Flanagan et al. (2010) suggested some specific adjustments to immunological assessment protocols in early life in low-income settings due to social confounding (Flanagan et al. 2010). These unmeasured confounding factors in vaccine response—such as non-specific effects and vaccine interactions, sex specific differences, nutritional status, and infection disease history—can alter the host immunity background. The authors suggested some practical considerations in longitudinal vaccine studies regarding data collection on potential confounding factors in infants such as breast feeding trends, vitamin or anthelminthic intervention in the population, and general health of the study population (Flanagan et al. 2010). In a meta-analysis of prospective studies of the efficacy of the BCG vaccine, Wilson, Fineberg, and Colditz (1995) show that geographic latitude is the source of variation in the results from trials and case-control studies, accounting for 41 percent of the between-study variance. The main differences across geographic locations are the social features of a population, such as its cultural and sociological factors (Wilson, Fineberg, and Colditz 1995). Therefore, to have a better understanding of the human response to vaccines, we should also take into consideration other variables that could influence the immune response, such as socio-economic status (SES), nutrition, and vaccination history.

Host immunity background responses vary in each population depending on the infection history. In the case of the BCG vaccine, though it is one of the most used vaccines, recent studies have shown that there is a geographical variation in its efficacy. It appears that exposure to environmental mycobacteria in LMICs provides a distinct immunity background for each population. Indeed, LMIC countries with higher TB prevalence are usually those in which the populations are constantly exposed to various microorganisms (Kaufmann, 2010). Cross-reactions of environmental mycobacteria-inducing T-helper 1 cell (immune cells mediating immunity against intracellular parasites, Th1) responses are believed to prevent the replication of the attenuated live bovine tuberculosis bacillus present in the vaccine. Such replication is needed to induce protection, and its absence causes the failure of the BCG vaccine (Kaufmann 2010). In addition, a recent study on human microbiome indicated that it is important to consider the composition of intestinal microbiota playing a role in oral-administered vaccine efficacy (Ferreira, Antunes, and Finlay 2010). Intestinal microbiota of a population in low-income countries are different than the ones in high-income countries due to their diets and living conditions. Thus, an oral vaccine could be processed differently depending of the variation in microbial population in their stomachs.

The season in which the vaccine was administered in a study in Pakistan and Gambia affected the serum antibody levels produced in the different population groups as a response to different vaccines (Moore et al. 2006). The difference in antibody levels was suggested to be due to the seasonal pattern of malaria transmission, which is usually higher in the rainy season (Moore et al. 2006). Malaria is a widely-spread parasitic infection in LMICs, and the immune responses towards vaccine antigens, as with the BCG vaccine, were found to be suppressed in heavily parasitized populations (Labeaud et al. 2009). Thus, exposure to parasites was closely related to vaccine failure.

Additionally, Labeaud et al. (2009) have noted that vaccine responses in children may also be influenced by maternal parasitic infections. Longitudinal maternal–infant cohort studies showed that an immunomodulatory phenotype was induced and continued into infancy, and later into childhood, due to the exposure to parasites in utero that altered protective responses to antigens from childhood vaccines (Labeaud et al. 2009). However, it was suggested that successful treatment of pregnant mothers for parasitic infections may improve vaccination outcomes of the unborn child. Furthermore, in the study of Cooper et al. (2001), it was found that adults infected by *Ascaris lumbricoides*, an extracellular parasitic worm, had lower Th1 immune response than the population that was not infected by the parasite. It was believed that parasite infections induced T-helper 2 cell (immune cells mediating immunity against extracellular parasites, Th2) responses, thus suppressing Th1 cells with results that change the immune responses to orally administered vaccines. For example, in an interview conducted in the present study, one of the public health experts illustrated this with examples from South Asia, where the efficacy of the Oral Polio Vaccine in Pakistan, Bangladesh, and India is less than in other parts of the world. It is suggested that the low efficacy of the vaccine is due to other enteric infections that influence the vaccine. Furthermore, the expert also pointed to a recent theory about the increasing asthma prevalence in developed countries as a result of reduced microbial burden during childhood (Senior Expert in Communicable Diseases, Stockholm, August 9, 2012).

Another important component of the immunity background is exposure to environmental mycobacteria, which are also associated with seasons

(Dockrell 2008). Seasonally-dependent environmental factors, other than infectious agents, are believed to play a role in vaccine-induced immune response (Moore et al. 2006). Such environmental agents are believed to act as an adjuvant to increase the antibody response at specific times of the year. Understanding the socioeconomic and environmental factors in the community, as well as the association with mechanisms of antibody production to vaccination, could be critical in the vaccine administration in specific population groups. Environ-vaccinology would argue to adapt vaccination schedules in countries with seasonally distinct environments for the most effective vaccination results.

Furthermore, different populations may have diverse vaccination histories because of the country-specific vaccination policy and sociocultural circumstances. The combination of different vaccinations in a population might cause different immune responses to certain vaccines (Shann 2000). Aaby and Benn (2011) demonstrated that the Oral Polio Vaccine (OPV) possibly has an effect on the response to the BCG vaccine (Sartono et al. 2010). The authors compared infants in Guinea-Bissau to whom the BCG and OPV vaccines were simultaneously administered with infants that only received BCG. The results showed that OPV suppressed the immune response induced by the BCG vaccine, with the result being more significant in low-birth-weight infants (Sartono et al. 2010). Prior vaccination history is therefore believed to affect the immune response of future vaccinations (Aaby and Benn 2011; Sartono et al. 2010).

The importance of vaccination history or vaccination order in influencing the vaccine response was also highlighted in an interview with an expert in communicable diseases. He argued that it is important to do research on the vaccination order because the public is usually hesitant to take many vaccines, or different vaccines, in a short time period (Senior Expert in Communicable Diseases, Stockholm, August 9, 2012). Therefore, it is helpful to have evidence to convince the public about the effectiveness of vaccines according to a vaccination schedule.

A range of sociocultural factors could affect prior vaccination history. Host immunity background and heterologous immunity that is acquired through different vaccines will require that public health experts in that specific area reconsider the order of vaccines that are being introduced to the population. Furthermore, taking into account the role of gender and sexual hormones differences is important in order to adjust vaccine doses or types, as the differences in vaccine responses between females and males has been suggested by many articles (Claesson 2011; Veirum et al. 2005). Women and men can also be exposed differently to parasitic diseases as a result of gender roles shaped by the society. In different parts of the world, gender exposes women to occupational risks, sociocultural behavior, and practices different than men (Okwa 2007). The vaccination history of children also depends on the beliefs and educational level of mothers; women who have higher educational levels and use prenatal care are more likely to vaccinate their children (Choi and Lee 2006). In addition, in settings in which gender discrimination is common, immunizations might differ depending on the child's sex (Choi and Lee 2006).

Beliefs and practices in different cultures can also affect what type of food is consumed, especially by children. These beliefs and practices are also factors influencing vaccine response. In a study by Silverfdal, Ekholm, and Bodin (2007), breast-feeding was found to enhance immune response to vaccines in children because it affects the intestinal microflora in infants. The study suggested that similar feeding trends should be considered in vaccine efficacy studies (Silfverdal, Ekholm, and Bodin 2007). In a different setting in Laos, food practices of lactating mothers usually decreased the variety of their children's food intake, and sick children were usually offered less foods and insufficient nutrients, causing a higher risk of malnutrition (Phengxay et al. 2007). Again, malnutrition and poor nutritional status are usually related to a poor immune system. Protein Energy Malnutrition (PEM) is known as a common cause of secondary immune deficiency in children (Taylor et al. 2012). The deficiency in the immune system might have an impact on the population vaccine response. In one study, the research showed that the PEM induced in mice by a low protein diet has a negative impact on the CD 8 memory, which can alter the vaccine response (Iyer et al. 2012). Socioeconomic status can affect nutrition intake in a household, and children's diet in rural areas of LMICs is often insufficient in quality and quantity (Irala-Estévez et al. 2000). An expert in communicable diseases who was interviewed mentioned the interaction between nutritional status and antibody response with regard to vaccination response. He argued that maternal nutrition and micronutrient deficiency can have an impact on vaccine response. Also, he emphasized that it would be helpful if we could evaluate if maternal nutrition or health have a direct role on the child's vaccine response, since most of the vaccines are administered very early in life (Senior Expert in Communicable Diseases, Stockholm, August 9, 2012). Similarly, an expert in vaccine preventable diseases argued that the immune response is also dependent on socioeconomic status (SES). If you have a good SES, you generally have better nutrition, and you are better prepared to fight diseases (Scientific Officer, August 25, 2012).

The effects of micronutrient deficiency such as vitamin A, zinc, and iron in vaccine response have drawn huge interest from researchers (Savy et al. 2009). The World Health Organization (WHO) has suggested that vitamin

A supplementation should be implemented in vaccination projects in countries that have high rates of vitamin A deficiency (WHO, 1998). However, there is a complex debate ongoing about the benefits of vitamin A in combination with vaccines. In a study by Benn et al., it was shown that vitamin A might have negative effects when introduced with the DTP vaccine (Benn et al. 2009, 2010). Moreover, recent meta-analysis literature showed that poor nutritional status has little or no effect on vaccine responses, and the benefit effects of micronutrients on vaccine response are also weak (Savy et al. 2009). The review also mentioned that poor quality and heterogeneity of the data makes it harder for the analysis to draw a firm conclusion. Further studies with detailed examination on immunological mechanisms are still needed to confirm the association between nutrition and vaccine response (Savy et al. 2009).

The environ-vaccinology approach is different from the basic immunological research conducted that predominantly considers only experimental variables and controls for other confounders. These so-called "controlled variables," such as nutritional and diet factors, coexisting diseases, and social surrounding, do exist in real life and may influence different immune pathways in our body. For instance, variation in host factors, such as stress, is usually treated as a confounder in immunological research. Interdisciplinary research between psychologists and immunologists showed that there is an influence of human stress in vaccine response (Phillips 2011). Each one of the molecules, cells, and tissues in the complex network of the immune system has a specific role; all of them interact with each other in a corresponding system to generate the immune response to a vaccine. If the social confounders play an important role in biological processes in real world setting, how can one technically include social differences in the local context into a scientific standard? The molecular mechanisms by which environmental or social factors can directly or indirectly influence the immune response are, at the moment, unknown. Environmental and social variation may have certain effects on the transcription and translation process of genes and may even influence protein functions responsible for different vaccine responses. For example, a high-fat diet (social factor) can induce chronic oxidation that inhibits mucosal immunity (Cui et al. 2009). With new advances in technology and in the field of systems biology, new insights into the interconnected networks that control and produce immune responses can be provided (Querec et al. 2009). It is important to incorporate multiple levels of knowledge on social and environmental variation information and pathways of interaction with host biological factors. Therefore, with the help of social scientists, a comprehensive understanding of sociocultural factors in a community can be gained in order to fill in the missing information in the linkage mapping of environ-vaccinology. Looking at the immune system as a whole, rather than focusing on one small part of it, represents an important shift of vaccine development.

Although it is difficult to determine to what degree vaccine responses are influenced by social factors, and where the relationships between confounding factors begin, it is evident from the literature review that social confounders can have an impact on the vaccine response. Vaccine development research has always put emphasis on basic research of biological components, but this tends to overlook the host environment and social surrounding interactions that contribute to the heterogeneity of immune responses. Future studies that are able to classify social heterogeneity of the population may provide the ability to predict interpopulation differences in vaccine responses. With such results, one will be able to tailor an effective vaccine according to the specific population, optimizing vaccine response by different vaccine developments and vaccination implementations.

## Perspectives of Public Health and Vaccine Experts on Environ-Vaccinology Innovations and Policy Translation

The following section highlights the perspectives of public health and vaccine experts about the environ-vaccinology theory and its possible policy applications for public health. Though the notion of environ-vaccinology was new for most of the respondents, they are well aware of socioenvironmental factors affecting vaccine response. The main themes that emerged from the interviews were the challenges in environ-vaccinology research design and the implementation of environ-vaccinology in public health policies.

### Challenges in Applying Environ-Vaccinology Research Design

The progress of translating new knowledge in basic scientific research into public health and health practice may be slow. Although several of the experts think that environ-vaccinology offers a worthy opportunity to enhance the vaccination programs, one also argued that it would be difficult for the vaccine experts in the field of public health to accept and apply the new insights. Despite available scientific evidence, political, social, legal, and economic factors may challenge application (Interview Scientific officer, August 19, 2012). Environ-vaccinology implies that the social variability between populations with increased challenges for research design and analysis should be taken seriously. For example, measuring the impact of disease exposure on vaccine response is difficult due to the variety of variables in each population. A public health expert in environmental determinants of health argued that:

*Clearly you know HIV ... would have a big impact on immune system, thus influence the vaccine response. But to be able to understand all the different spectrum of infectious diseases that people have in a man and how those might impact the vaccine response, could be quite complicated and difficult, because everyone could be expected to be slightly different in their disease history ... or let us's just say bacteria and viruses which are not causing diseases at the moment, so that would be quite difficult.* (Expert in Environmental and Social Determinants of Health, Stockholm, August 10, 2012)

He also emphasized that it is unethical to run an experiment that tests the population under different vaccine or vaccination schedules. Therefore, it would be a challenge to understand what vaccine, and under what schedule, will be the best for each population.

Another challenge could lie in the implementation of the scientific insights gained through environ-vaccinology research in policy and established research and development practices. This often requires engagement with key stakeholders across many levels, as well as high-level government commitment. Due to the conservative mind-set of vaccination programs, experts envision difficulties in introducing the environ-vaccinology to key stakeholders in vaccine development. According to the respondents, vaccine development is a conservative field that has not changed much since the first vaccine was introduced. This is mostly due to the increasing concerns regarding safety of vaccines, along with a rise of antivaccination movements recently. A respondent stated:

*I think it will be really hard to give the new ideas to the public health officials particularly in the area of vaccination, those who work in the vaccination program are quite conservative.* (Senior Expert in Communicable Diseases, Stockholm, August 9, 2012)

*Usually we are very conservative about how we move in that [vaccinology] field, because you need to offer the person something, that it's going to be safe, that means that he or she is not going to be worse after taking the vaccine than before. You need to ensure that they are going to be protected and it's like an investment in the future. So it means efficacy, and it means also immunogenicity. That's many things you need to assure and many factors. I think it's difficult to move to something else unless you prove scientifically that the mechanism works and has other values.* (Expert in Vaccine-Preventable Diseases, September 18, 2012)

Similarly, another expert, who has worked in the public health system for years, explained the difficulties in changing the standard vaccine development among vaccine manufacturers. Significant challenges in changing the vaccine development were investments in new vaccine development protocol, the safety of the newly developed vaccines, and also the conservativeness of the vaccine experts. Therefore, most of the people are not motivated to change the current process:

> *Anyone who is comfortable to the current situation would not have much incentive to make their life more difficult, the big vaccine manufacturer for example. There might be individuals or small unit who might think it is interesting to explore but they are probably not going to lead the way on this topic until there is evidence-based. Even so, the people involved [public health system, health services, and vaccine suppliers] know that changing is so difficult, you will never overcome a significant barrier and it involves a huge amount of change.* (Expert in Environmental and Social Determinants, July 15, 2012)

He further described how barriers (e.g., absence of funding) could stop manufacturers from considering the application of environ-vaccinology. Also, the need for policy and structural change would be extremely difficult to alter in the public system. This can be due to different challenges, such as the difficulties to convince public health officials without detailed, evidence-based results and translating science innovation into public health. He gave the example of harmonization of the vaccine schedule across multiple countries being very difficult to implement due to political and policy issues. According to him, it will be more complicated to draft a vaccine protocol at the community level. This will need a multisectoral approach and high-level government commitment to redesign and develop new protocols and guidelines that ensure coverage of different communities.

### Challenges in Implementing Environ-Vaccinology Innovations in Public Health Policies

New scientific knowledge can change regulatory thinking of decision makers and challenge stakeholders to design new guidelines overseeing the development and use of vaccinations. Environ-vaccinology suggests that to ensure the appropriate vaccine doses and schedule are suited to a specific population in the right season of the year, vaccine policies and protocols will need to be adapted to the local context of environmental factors. This means that vaccine protocols should be drafted at the community level rather than the country or regional level, since environmental factors may vary among communities. However, critical sociocultural data for decision making about vaccine types and schedules are often incomplete or entirely lacking.

All experts interviewed agreed that to encourage the change of the paradigm and conceptual framework of vaccine development, it is necessary to get the attention of funding and public health agencies and create more evidence. However, attracting funding or public health agencies to support a new research field is challenging. It takes a significant amount of time and considerable effort to create a critical mass of research publication. One of the respondents hoped that, with the growing attention and public funding for innovative research collaborations, there will also be more support to encourage more researchers to create evidence on environ-vaccinology (Expert in Environmental and Social Determinants of Health, Stockholm, August 10, 2012). Another expert highlighted the importance of evidence for differences among large populations in order to generate funding. If, for instance, research evidence showed that there is a big difference of rotavirus vaccine efficiency in South America compared to North America, vaccine manufactures would probably be interested because it affects a large population. Therefore, if researchers can create a system to assess the impact of socio-environmental factors on vaccine response, the vaccine manufacturers will follow suit. Several experts also mentioned that vaccine manufacturers should act as financial sponsors of the research, again stressing the role of evidence-based research in this field to facilitate such commercialization:

> It should be important to consider vaccine manufacturers. (Expert in Vaccine-Preventable Diseases, September 18, 2012)

> Vaccine development is a business in the end, so most of the money is coming from the industry. (Scientific Officer, August 25, 2012)

> You need some evidence that there are vaccines that are working differently in different population. Otherwise, I don't think the vaccine manufacturer in the preliminary phases of the vaccine development will investigate this. (Senior Expert in Vaccine-Preventable Diseases, July 15, 2012)

Similarly, one respondent mentioned the example of the HPV vaccine, which was discovered and distributed in a very short amount of time because of the high level of support and investment from vaccine manufacturers:

> The development of the HPV vaccine is phenomenal. It's incredibly fast. I think this is unprecedented in history. From the time they discover the link between infection to the cancer, and they develop the vaccine and the uptake is incredibly fast. And vaccine manufacturers play a big role in this because they see the market. (Senior Expert in Communicable Diseases, Stockholm, August 9, 2012)

According to him, if there is little commercial value, it is difficult to get support from the drug companies. There are only a few companies that remain in the complex vaccine business of developing and producing vaccines. Besides the science of developing a vaccine, the manufacturing process is long and expensive. If there is lack of commercial value in creating certain types of vaccines targeting specific population and regions, the drug company will not be motivated to support the research.

Generally, the experts expressed that investing in the new field of vaccinology can be uncertain, with significant risks involved. The vaccine business is expensive; it accesses a smaller market compared to that of, for example, pharmaceuticals. The experts saw important challenges in implementing the new scientific knowledge in vaccine development. These include engaging multidisciplinary groups of actors, such as scientists from life sciences and social sciences, decision makers, public health officers, and community members. Furthermore, attracting and securing support from funding agencies and vaccine manufacturers is another obstacles in environ-vaccinology research. Overall, the experts were in agreement with the idea of expanding the scientific understanding of how the social environment affects vaccine response. They also agreed that it would be helpful to introduce changes into the rather conservative field of vaccination, while supporting public safety concerns about vaccines, and potentially generating financial support for more research from industry.

## Discussion and Conclusion

It is now possible to vaccinate against various types of deadly infectious diseases. However, the outcome of vaccinations is affected by the conditions that alter host immunity background such as disease history, nutritional status, socioeconomic status, vaccination history, sex and gender of a population, and complicated sequences of vaccines, which are especially common in low-income settings (Hernandez and Blazer 2006). Thus far, researchers have not paid sufficient attention to the potential interactions between vaccine response and socioenvironmental factors that are likely to contribute to the preventive effect of vaccines in most populations. As such, it is the common assumption that universal, "one size fits all" vaccines are efficient for different populations around the world. The literature reviewed in this chapter shows that socioenvironmental factors affect vaccine efficacy of different populations in important ways. In vaccine development studies, the central issue is population-to-population and person-to-person variation in producing a protective immune response. Traditional approaches for vaccine development and vaccine trials often assume that vaccine efficacy is homogeneous for different population across the world. However, variation of socioenvironmental factors may influence a vaccine's protective effect. This may imply that high vaccination coverage might not induce mass immunity of a population due to the failure of the vaccine. Developing vaccines without taking into consideration the social confounders of a population could thus affect the vaccine efficacy. Socioenvironmental factors and vaccines should therefore be seen not only as independent effects, but also as immuno-modulators, affecting the overall mass immunity of a population.

From this literature review, it follows that the broader social context within a population cannot be ignored during the development of vaccine protocols. Different communities have distinct immunity backgrounds and might need different vaccine types or doses. Insights into the local context of socioenvironmental factors might furthermore support better design of vaccine programs. This could be accomplished by adapting vaccine dosages and schedules to specific populations and seasons of the year. Drafting vaccine protocols at the community level rather than at the country or regional level might be necessary, since the environmental factors may vary among communities.

However, the literature review also showed that critical sociocultural data for making these kinds of decisions are often incomplete or entirely lacking. To provide the potential benefit of environ-vaccinology to public health, a multidisciplinary approach and political will to collect such data must be considered. The results from the interviews with public health and vaccine experts highlighted the need for more research into this field in order to gain the support of funders and public health officers. Future research is considered necessary to investigate the relationship between vaccine response and socioenvironmental factors for different contexts, populations, and environments.

Based on the literature and the interviews with public health and vaccine experts, it seems that the key to the success of this research is to work in collaboration with multidisciplinary groups, including social scientists, and generate predictive models of vaccine responses at the individual, gender-specific, or sub-population level. While using one universal vaccine for the entire world population can be convenient and can lead to the prevention of diseases in certain populations, it can also provide little benefit for public health while wasting resources. Thus, to optimize the vaccine efficacy, the new field of environ-vaccinology suggests examining the effects of different social confounders in vaccine response and their potential interactions with other pathways in immunology. Further evidence-based research in this field can support the design, development, and production of next-generation vaccines and help improve their immunological response. Environ-vaccinology thereby offers the means to adapt a universal solution, such as a vaccine, to different local contexts. By adapting vaccines to individual bodies and local biosocial contexts, environ-vaccinology innovations promise to circumvent the tension between universal standards and local adaptation.

## Acknowledgements

The author would like to acknowledge the help of Professor Daniel D. Reidpath from Monash University for the suggestion of the topic and ECDC experts for the interview time and insights provided.

## References

Aaby, P., and C. S. Benn. 2011. "Non-Specific and Sex-Differential Effects of Routine Vaccines: What Evidence Is Needed to Take These Effects into Consideration in Low-Income Countries?" *Human Vaccines* 7(1): 120–124.

Azad, M. B., Y. Lissitsyn, G. E. Miller, A. B. Becker, K. T. HayGlass, and A. L. Kozyrskyj. 2012. "Influence of Socioeconomic Status Trajectories on Innate Immune Responsiveness in Children." *PLoS ONE* 7(6): e38669. doi:10.1371/journal.pone.0038669.

Benn, C. S., A. B. Fisker, B. M. Napirna, A. Roth, B. R. Diness, K. R. Lausch, H. Ravn, et al. 2010. "Vitamin A Supplementation and BCG Vaccination at Birth in Low Birthweight Neonates: Two by Two Factorial Randomized Controlled Trial." *BMJ (Clinical research ed.)* 340: c1101.

Benn, C. S., A. Rodrigues, M. Yazdanbakhsh, A. B. Fisker, H. Ravn, H. Whittle, and P. Aaby. 2009. "The Effect of High-Dose Vitamin A Supplementation Administered with BCG Vaccine at Birth May Be Modified by Subsequent DTP Vaccination." *Vaccine* 27(21) : 2891–2898. doi:10.1016/j.vaccine.2009.02.080.

Bernstein, A., B. Pulendran, and R. Rappuoli. 2011. "Systems Vaccinomics: The Road Ahead for Vaccinology." *OMICS: A Journal of Integrative Biology* 15(9): 529–531. doi:10.1089/omi.2011.0022.

Black, G. F., R. E. Weir, S. Floyd, L. Bliss, D. K. Warndorff, A. C. Crampin, B. Ngwira, et al. 2002. "BCG-Induced Increase in Interferon-Gamma Response to Mycobacterial Antigens and Efficacy of BCG Vaccination in Malawi and the UK: Two Randomized Controlled Studies." *Lancet* 359(9315): 1393–1401. doi:10.1016/S0140-6736(02)08353-8.

Centers for Disease Control and Prevention (CDC). (1999). Impact of vaccines universally recommended for children—United States, 1990–1998. *MMWR. Morbidity and Mortality Weekly Report*, 48(12), 243–248.

Choi, J. Y., and S. H. Lee. 2006. "Does Prenatal Care Increase Access to Child Immunization? Gender Bias among Children in India." *Social Science and Medicine (1982)* 63(1): 107–117. doi:10.1016/j.socscimed.2005.11.063.

Claesson, M. H. 2011. "Immunological Links to Nonspecific Effects of DTwP and BCG Vaccines on Infant Mortality." *Journal of Tropical Medicine* 2011. doi:10.1155/2011/706304.

Cohen, S., and T. B. Herbert. 1996. "Health Psychology: Psychological Factors and Physical Disease from the Perspective of Human Psychoneuroimmunology." *Annual Review of Psychology* 47(1): 113–142. doi:10.1146/annurev.psych.47.1.113.

Cooper, P. J., M. Chico, C. Sandoval, I. Espinel, A. Guevara, M. M. Levine, G. E. Griffin, T. B. Nutman. 2001. "Human Infection with Ascaris Lumbricoides

Is Associated with Suppression of the Interleukin-2 Response to Recombinant Cholera Toxin B Subunit Following Vaccination with the Live Oral Cholera Vaccine CVD 103-HgR." *Infection and Immunity* 69(3): 1574–1580. doi:10.1128/IAI.69.3.1574-1580.2001.

Cui, J., G. Le, R. Yang, and Y. Shi. 2009. "Lipoic Acid Attenuates High Fat Diet-Induced Chronic Oxidative Stress and Immunosuppression in Mice Jejunum: A Microarray Analysis. *Cellular Immunology* 260(1): 44–50. doi:10.1016/j.cellimm.2009.08.001.

Dockrell, H. M. 2008. "Real Vaccines in the Real World: Tuberculosis Vaccines Move South." *Expert Review of Vaccines* 7(6): 703–707. doi:10.1586/14760584.7.6.703.

Dowd, J. B., M. N. Haan, L. Blythe, K. Moore, and A. E. Aiello. 2008. "Socioeconomic Gradients in Immune Response to Latent Infection." *American Journal of Epidemiology* 167(1): 112–120. doi:10.1093/aje/kwm247.

Ehreth, J. 2003. "The Global Value of Vaccination." *Vaccine* 21: 596–600.

Ferreira, R. B. R., L. C. M. Antunes, and B. B. Finlay. 2010. "Should the Human Microbiome Be Considered When Developing Vaccines?" *PLoS Pathog* 6(11): e1001190. doi:10.1371/journal.ppat.1001190.

Flanagan, K. L., S. Burl, B. L. Lohman-Payne, and M. Plebanski. 2010. "The Challenge of Assessing Infant Vaccine Responses in Resource-Poor Settings." *Expert Review of Vaccines* 9(6): 665–674. doi:10.1586/erv.10.41.

Flanagan, K. L., R. van Crevel, N. Curtis, F. Shann, and O. Levy. 2013. "Heterologous ('Nonspecific') and Sex-Differential Effects of Vaccines: Epidemiology, Clinical Trials, and Emerging Immunologic Mechanisms." *Clinical Infectious Diseases* 57(2): 283–289. doi:10.1093/cid/cit209.

Hernandez, L. M. and D. G. Blazer, eds. 2006. *Genes, Behavior, and the Social Environment: Moving Beyond the Nature/Nurture Debate*. Institute of Medicine (US) Committee on Assessing Interactions among Social, Behavioral, and Genetic Factors in Health. Washington (DC): National Academies Press (US). http://www.ncbi.nlm.nih.gov/books/NBK19929/. August 13, 2013.

Irala-Estévez, J. D., M. Groth, L. Johansson, U. Oltersdorf, R. Prättälä, and M. A. Martínez-González. 2000. "A Systematic Review of Socioeconomic Differences in Food Habits in Europe: Consumption of Fruit and Vegetables." *European Journal of Clinical Nutrition* 54(9): 706–714.

Iyer, S. S., J. H. Chatraw, W. G. Tan, E. J. Wherry, T. C. Becker, R. Ahmed, and A. F. Kapasi. 2012. "Protein Energy Malnutrition Impairs Homeostatic Proliferation of Memory CD8 T Cells." *Journal of Immunology (Baltimore, Md.: 1950)* 188(1): 77–84. doi:10.4049/jimmunol.1004027.

Kaufmann, S. H. E. 2010. "Future Vaccination Strategies against Tuberculosis: Thinking Outside the Box." *Immunity* 33(4): 567–577. doi:10.1016/j.immuni.2010.09.015.

Kennedy, R. B., and G. A. Poland. 2011. "The Top Five 'Game Changers' in Vaccinology: Toward Rational and Directed Vaccine Development." *OMICS: A Journal of Integrative Biology* 15(9): 533–537. doi:10.1089/omi.2011.0012.

Labeaud, A. D., I. Malhotra, M. J. King, C. L. King, and C. H. King. 2009. "Do Antenatal Parasite Infections Devalue Childhood Vaccination?" *PLoS Neglected Tropical Diseases* 3(5): e442. doi:10.1371/journal.pntd.0000442.

Moore, S. E., A. C. Collinson, A. J. C. Fulford, F. Jalil, C. A. Siegrist, D. Goldblatt, L. A. Hanson, A. M. Prentice. 2006. "Effect of Month of Vaccine Administration on Antibody Responses in The Gambia and Pakistan." *Tropical Medicine and International Health: TM and IH* 11(10): 1529–1541. doi:10.1111/j.1365-3156.2006.01700.x.

Okwa, O. O. 2007. "Tropical Parasitic Diseases and Women." *Annals of African Medicine* 6(4): 157–163.

Phengxay, M., M. Ali, F. Yagyu, P. Soulivanh, C. Kuroiwa, and H. Ushijima. 2007. "Risk Factors for Protein-Energy Malnutrition in Children under Five Years: Study from Luangprabang Province, Laos." *Pediatrics International: Official Journal of the Japan Pediatric Society* 49(2): 260–265. doi:10.1111/j.1442-200X.2007.02354.x.

Phillips, A. C. 2011. "Psychosocial Influences on Vaccine Responses." *Social and Personality Psychology Compass* 5(9): 621–633. doi:10.1111/j.1751-9004.2011.00378.x.

Poland, G. A., I. G. Ovsyannikova, and R. M. Jacobson. 2008. "Vaccine Immunogenetics: Bedside to Bench to Population." *Vaccine* 26(49): 6183–6188. doi:10.1016/j.vaccine.2008.06.057.

Ponton, F., K. Wilson, S. C. Cotter, D. Raubenheimer, and S. J. Simpson. 2011. "Nutritional Immunology: A Multi-Dimensional Approach." *PLoS Pathog* 7(12): e1002223. doi:10.1371/journal.ppat.1002223.

Querec, T. D., R. S. Akondy, E. K. Lee, W. Cao, H. I. Nakaya, D. Teuwen, A. Pirani, et al. 2009. "Systems Biology Approach Predicts Immunogenicity of the Yellow Fever Vaccine in Humans." *Nature Immunology* 10(1): 116–125. doi:10.1038/ni.1688.

Sartono, E., I. M. Lisse, E. M. Terveer, P. J. M. van de Sande, H. Whittle, A. B. Fisker, and C. S. Benn. 2010. "Oral Polio Vaccine Influences the Immune Response to BCG Vaccination. A Natural Experiment." *PLoS ONE* 5(5): e10328. doi:10.1371/journal.pone.0010328.

Savy, M., K. Edmond, P. E. M. Fine, A. Hall, B. J. Hennig, S. E. Moore, K. Mulholland, U. Schaible, A. M. Prentice. 2009. "Landscape Analysis of Interactions between Nutrition and Vaccine Responses in Children." *The Journal of Nutrition* 139(11): 2154S–2218S. doi:10.3945/jn.109.105312.

Shann, F. 2000. "Non-Specific Effects of Vaccines in Developing Countries." *BMJ: British Medical Journal* 321(7274): 1423–1424.

Silfverdal, S. A., L. Ekholm, and L. Bodin. 2007. "Breastfeeding Enhances the Antibody Response to Hib and Pneumococcal Serotype 6B and 14 after Vaccination with Conjugate Vaccines." *Vaccine* 25(8): 1497–1502. doi:10.1016/j.vaccine.2006.10.025.

Taylor, A. K., W. Cao, K. P. Vora, J. D. L. Cruz, W.-J. Shieh, S. R. Zaki, S. Sambhara, S. Gangappa. 2012. "Protein Energy Malnutrition Decreases Immunity and Increases Susceptibility to Influenza Infection in Mice." *Journal of Infectious Diseases* jis527. doi:10.1093/infdis/jis527.

Veirum, J. E., M. Sodemann, S. Biai, M. Jakobsen, M.-L. Garly, K. Hedegaard, H. Jensen, P. Aabyl. 2005. "Routine Vaccinations Associated with Divergent Effects on Female and Male Mortality at the Pediatric Ward in Bissau, Guinea-Bissau." *Vaccine* 23(9): 1197–1204. doi:10.1016/j.vaccine.2004.02.053.

WHO, UNICEF. Integration of vitamin A supplementation with immunization: policy and programme implications.WHO/EPI/GEN/98.07. WHO. Geneva. 1998

Wilson, M. E., H. V. Fineberg, and G. A. Colditz. 1995. "Geographic Latitude and the Efficacy of Bacillus Calmette-Guérin Vaccine." *Clinical Infectious Diseases* 20(4): 982–991. doi:10.2307/4458467*Human vaccines*, 7(1), 120–124.

# CHAPTER 8

# Different Perspectives on Research and Development Incentives for Diseases of the Poor

*Lois A. Murray and David Townend*

## New Medicines for Diseases of the Poor

In the field of global health, we are working toward a world in which everyone can exercise their fundamental "right to the enjoyment of the highest attainable standard of physical and mental health" (World Health Organization 1946). This encompasses the right to access to medicines, including the availability of new medicines (UN General Assembly 2008). Research and development (R&D) for new medicines should therefore meet the health needs of all people. Unfortunately, current R&D efforts largely neglect diseases that primarily affect the poor.[1] Out of the 1,035 drugs developed between 1975 and 1999, only one percent of these were for diseases that disproportionally affect the poor (Hubbard and Love 2004). This lack of effective treatments is leaving millions of people worldwide at risk of disability and death.

The divergence between health needs and the current direction of R&D is a matter of concern for the global community. Globalization has facilitated the international spread of people, food, and infectious diseases at an unprecedented rate. One example is the epidemic of Severe Acute Respiratory Syndrome (SARS) that began in Hong Kong in 2002 and spread to 37 countries within weeks (Smith 2006). This outbreak illustrates the pressing need for integrated international measures in effective treatment of diseases. In this increasingly interconnected world, it is imperative that we take measures to address the gap between the development of new medicines and the global burden of disease. Diseases of the poor can no longer be regarded as a foreign concern.

According to Hotez et al. (2007), neglected diseases represent the sixth largest global health burden, yet there has been little progress in the R&D pipeline for medicines for these diseases (Trouiller et al. 2002). The dry pipeline is often attributed to the major influence of the private pharmaceutical industry in determining the direction of investment. However, pharmaceutical companies are actors working within the current framework of patent protection as an incentive for R&D. This incentive needs to be examined to determine if any change or additional incentives could help restore the balance of R&D. One potential alternative incentive is the use of "prizes" to stimulate innovation. The "prize" model that we are considering in this study is not our idea; indeed we are not contributing to the technical development of the "prize" model. It has been developed recently, particularly in the work of Hubbard and Love. We wished to consider it because it is an exciting development, both of itself and as an attempt to change a legal paradigm. Our study concerns how a new regulatory technology might be perceived by those who are stakeholders in making a change to the regulatory system.

There are several forms of R&D incentives. This study focuses on patents and prizes and explores each concept from a range of perspectives, with the aim of gaining important insights on the likely feasibility of implementing a prize system to reimburse and incentivize R&D. The R&D process consists of several stages that develop basic research discoveries into a useful and ideally accessible product. This is a risky, lengthy, and costly process; patents are currently the main incentive used to stimulate actors to engage in the process. There are many different actors inside the complex wheel of R&D. This study focuses on nine key "perspectives" that are visible in the area: administrative, civil society, governance, government, global health, human rights, legal, multinational pharmaceutical, and start-up innovator.[2]

The nine perspectives were identified through a literature review. Each perspective is involved in the R&D process in a unique way. The private sector will be at the core of any change, and acceptability of these perspectives is an important determinant of success. Public health and civil society were selected as perspectives from the demand side of the dilemma, as patients' needs are likely to be within their priorities. The human rights perspective was chosen to place R&D into the wider context of universal rights. Political will is needed to implement any change to the R&D framework; therefore, a government perspective was selected. Intellectual Property Management (IPM) is a contentious issue in R&D, as some argue that patent protection is a basic right of the inventor while others feel that it impedes access. Therefore, an intellectual property legal representative was chosen to discuss the role of IPM. Finally, a feasible alternative R&D incentive requires an effective administrative and governance system; these perspectives were chosen to provide insights into practicalities of an alternative incentive system (see Table 8.1).

The setting of the study was not defined by location; this is a study about the concept of prizes, which could be applied nationally, regionally, or globally in the context of R&D framework. It is argued that as prizes are not defined by geographical boundaries, neither should the research setting be define by such boundaries. The target group included all relevant key informants who were identified, through the literature review, as having a distinct perspective on prizes, regardless of the geographic location. One expert interviewee was obtained through purposive sampling for each perspective. The

**Table 8.1**  Purposive sampling of interviewees

| PERSPECTIVES | SOURCES OF INFORMATION | AREA OF OCCUPATION OF INTERVIEWEE |
|---|---|---|
| Administrative | Administrative organisation in area of intellectual property management | Pharmaceutical Division of Regional IPM office |
| Civil Society | Relevant Non-Governmental Organisation | Senior health & HIV policy advisor of key NGO |
| Governance | Pharmaceutical Regulatory Organisation | Director of clinical regulatory institute's International Programme |
| Government* (Pro Prizes) | Member of Parliament | Member of European Parliament. (Prize Advocate) |
| Human Rights | Human Rights Professional | Human Rights PhD candidate with focus on Access to Medicines |
| Intellectual Property/Legal | Intellectual Property Law Academic | Assistant Professor International Economic Law and Intellectual Property |
| Large-scale Private Sector | Multinational Pharmaceutical Company | Director of Developing Countries and Market Access unit at one of world's leading pharmaceutical companies |
| Global Health | Global Health Consultant | Consultant with expertise in Innovation and Public Health |
| Small-Scale Private Sector | Start-up Innovator Company | Founder of Start-up Biotech company |

purpose of this method was to include "information-rich cases for in-depth study" (Patton 1980). This method is not representative, but aims to provide insights from information-rich sources of each perspective.

Data was collected between May and August 2011 in the form of interviews, with one expert informant for each perspective. The first author conducted semi-structured, open-ended interviews, which were audio-taped and transcribed. The interviews were based on discovering the attitudes of the interviewees towards the current R&D incentive mechanism of patent protection and one alternative funding mechanism: prizes.

Semi-structured interviews were appropriate because they enabled the interviewer to focus the interview on the key issues while still leaving room for interviewees to expand on specific themes. Given the practical constraints of this research, telephone interviews were chosen as the most feasible method. Out of the nine interviews, seven were telephone interviews, one was face-to-face, and one was completed using Skype™. Variation in methods of data collection between interviews was minimized as much as possible by using the same researcher and standardized interview questions.

Areas explored in the interviews included the current incentive of patent protection for R&D, potential limitations, "delinking" R&D costs from prices of medicines, and the role of prizes as an alternative. This study is not representative of all opinions on R&D. It aims to present useful insights that can be used to stimulate further research and discussion in the area of alternative incentives for R&D.

## Innovation in Research and Development of Medicines

In order to understand the complex issues surrounding alternative incentives for innovation, we first need to examine the current innovation process for new medicines. The issue at hand is how the context of regulation, currently designed to incentivize R&D in relation to a particular pharmaceutical or disease sector that is widely regarded as failing that sector, can accommodate a new regulatory technology—the prize—or at least evaluate its potential and consider its implementation.

Traditionally, the R&D process has been depicted as following a linear trajectory from discovery to development, made up of consequential steps. Each step presents opportunities for failure along the process; this results in a complex pathway of substantial risk, length, and cost. However, the linear model has been criticized as failing to capture the increasing interactions between each stage of the process and the many different relationships within the R&D structure (Hirsch-Kreinsen and Jacobson 2008). Subsequently, the World Health Organization (WHO) Commission of Intellectual Property,

Innovation, and Public Health (CIPIH) created a new image of medical innovation—the Innovation Cycle, which acknowledges that innovation does not always follow one direction.

The CIPIH 2006 Report found that the Innovation Cycle works reasonably well in developed countries where effective demand and population needs are similar. However, the cycle does not function optimally in developing countries in which, despite a great need for new products, there is a lack of a demand because of people's limited purchasing power (WHO 2006). This results in little invention for diseases specific to developing countries because there are no incentives to encourage actors to engage in the costly and risky R&D process. The Association of the British Pharmaceutical Industry (ABPI) has reported that only one in five thousand compounds that are initially tested pass every stage of the clinical process to gain approval for use on people (ABPI 2011). Pharmaceutical industry estimates report that, for a single medicine, it costs an average of $800 million and takes between 10 and 12 years to complete development (DiMasi, Hansen, and Grabowski 2003). Some collaborative not-for-profit efforts have developed improved medicines at significantly lower costs. The Drugs for Neglected Diseases Initiative (DNDi) is a product development partnership that has created a portfolio of improved medicines over the past eight years with $100 million (DNDi 2011). Despite the disparity in estimated costs, it is undeniable that, due to the nature of a health product, the need for high quality control and extensive clinical trials means that the R&D process is always going to require substantial time and investment. This leads to the dilemma of determining an appropriate incentive or reward.

## Patents and Prizes

Patent protection is the current main form of incentive for R&D of medicines. Patents link R&D costs to the price of medicines by granting the inventor an exclusive right to sell that product for a limited time period, typically 20 years. This marketing monopoly means that the inventor can set the price of this medication, and the resultant profits are the main way that developers recover the costs of research and development (Salazar 1998).

Patents have a long-established role in pharmaceutical R&D; the first patented medicines originated in England in the seventeenth century; the oldest known patented medicine, "Anderson's Scots Pills," was patented in the 1630s (Griffenhagen and Young 2009). The length and extent of the use of patents in medicine indicates that it is a system that has successfully promoted innovation and is well understood by those in the field.

Prizes have also been used as an incentive for innovation for centuries. A famous example is the British Longitude Prize, won in 1773 by John Harrison for his revolutionary navigation system (KEI 2008). However, despite their success in other fields, the use of prizes in medical R&D is quite limited. Prizes are a form of "pull" incentive for R&D. "Pull" incentives are direct rewards for the successful achievement of objectives (WHO 2006). Prizes can either be offered for reaching interim steps along the R&D process—"milestone prizes"—or offered at the end of product development—"end prizes (WHO 2012)."

The idea of using prizes for medical innovation was primarily (but not exclusively) proposed by Jamie Love, an economist and founder of the non-governmental organization (NGO) Knowledge Ecology International (KEI), and Tim Hubbard, an accomplished researcher in the field of genomics. They made their first proposal for a new incentive system for R&D in 2004, promoting delinking, which is the phenomenon of separating the R&D costs from the price of the medicines (Hubbard and Love 2004).

Delinking is the conceptual basis of prizes. While patents are based on tying the cost and the price inextricably together, prizes separate the development phase from production phase by offering a cash prize instead of a patent, with the condition that generic production is allowed. The private sector commonly argues that there is a need to charge high prices for patented medication in order to recover R&D costs and ensure further investment into R&D. However, figures from the pharmaceutical industry itself show that R&D costs are only eight percent of global sales, which puts this rationalization into question (Love 2011). It is interesting that the private sector, through this common justification for the need for high prices to recover R&D costs, is effectively pointing out a major problem with patents: the "linking" characteristic.

Subsequently, one of the main benefits of prizes as an incentive for R&D is the underlying principle of delinking. Delinking has become a buzz word in the field of R&D and global health. NGOs such as Oxfam advocate for this approach in order to increase accessibility of medicines to the world's poor (Oxfam International 2008). However, this is not merely a case of social justice; there is an undeniable economic logic behind the idea of delinking a very costly one-time development process from much less expensive, repeatable production costs that have been endorsed by economists and health advocates alike (Stiglitz 2007).

The potential of prizes as an alternative R&D incentive has been recently recognized by a WHO Consultative Expert Working Group (CEWG) on Research and Development: Financing and Coordination. In 2010, prizes were classified as "unfeasible" by a WHO Expert Working Group (EWG)

(WHO 2010); this met with criticism, particularly from nongovernmental organizations who complained about the conflict of interest of some members of the EWG (KEI 2010). Consequently, the CEWG was created in 2010 to revise the analysis. At a presentation about the group's progress, prizes were categorized under "proposals that meet many of the CEWG criteria" (2011). The final report highlighted the importance of delinking (WHO 2012). This significant advance reflects more positive thinking toward prizes. However, prizes are still a new concept; therefore, the opinion of nine key stakeholders on the relative advantages or disadvantages of prizes compared to patent protection were explored.

## Do We Need to Protect the Place of Patents in the System?

*The patent is still the most used form of reward.* (government interviewee, June 2011)

Patent protection is undoubtedly the best known form of incentive for R&D of medicines. The familiarity of each interviewee with patent protection is a testament to this. Several respondents felt that in certain areas, the patent system is very effective in stimulating innovation. The main advantage of patent protection was identified to be that it is well understood by all of those working in and around it.

*I mean whatever the defects of the patent system, it is not exactly simple but it works in a way that businesses and other actors understand.* (global health interviewee, August 2011)

Several interviewees argued for the need for the existing system of patent protection. The start-up innovator suggested that the current R&D pathway makes it impossible to compete with other innovators without having patent protection, as it is a prerequisite for attracting investors and is needed to mitigate risk.

*Ultimately patents are a necessity or a fundament of starting up a business. If I don't have patents or rights to patents … if I don't have that, I typically will have a very difficult time finding an investor to put any money in. The simple reasoning is if the company fails to reach the final result, a product that is sold for big money, then the investor still has this 'collateral' a 'fundament' something that they can still sell off … It's the way the industry is currently funded. People will only invest in my company if I have rights, to do this. I mean if anybody around the corner can look at what I do and as soon as we publish any article, they can copy it, nobody will put the money in my company.* (start-up innovator interviewee, June 2011)

The administrative representative highlighted the argument commonly used by the private sector, which is that patents are needed to compensate for the costly R&D process. The cost of R&D was also emphasized by both representatives of the private sector.

*"We have the view the patent for primary research is justified ... It is a fair reward because it is very costly to develop new drugs."* (administrative interviewee, June 2011)

The multinational pharmaceutical industry representative offered two pieces of evidence for the effectiveness of patent protection: the investment of the industry into R&D and the successful production of medicines every year:

*So I think my view is that the system is effective. And I think that there are two pieces of evidence for that effectiveness, the first is that the industry, pharmaceutical industry, invests about 100 billion dollars a year in R&D and the second piece of evidence is that the pharmaceutical industry produces roughly 20 medicines a year as a result of that investment ... it clearly incentivizes R&D, you wouldn't have those numbers there if it wasn't incentivizing.* (multinational pharmaceutical interviewee, July 2011)

It appears that, in the opinion of the interviewees, patents do stimulate innovation in some areas; from some perspectives, patents are an essential part of R&D as we currently understand it. The view of patents as a necessary part of the innovation process is not just confined to the interviewees but is reflected in international agreements on intellectual property management, such as the World Trade Organization Agreement on Trade Related Aspects of Intellectual Property Rights (TRIPS). Arguably, the familiarity of governments, the private sector, and public researchers with patents is one of the main reasons why the discussion of alternative incentives is so difficult. There was a feeling from the discussions that the patent incentive paradigm is so firmly rooted in innovation culture that it is difficult to imagine different paradigms. Nonetheless, one of the key findings of the study is that, despite the varying levels of support for patents, the interviewees acknowledged that there may be specific limitations to patent protection as an incentive for R&D of medicines for diseases of the poor.

## Limitations of Patents

The main argument used to highlight the limitations of patent protection was market failure. Market failure can be defined as "the inability of markets to reflect the full social costs or benefits of a good, service, or state of the world. Therefore, markets will not result in the most efficient or beneficial allocation of resources" (Ecosystem Valuation 2011). The failure in developing countries is that the people/countries in need of medicines are unable to pay for them; therefore, there is no effective market. As the patent protection mechanisms work by rewarding a market monopoly, there is a failure because there is no incentive for developers to make these medicines.

> *Well I think that there is a major problem in aligning the priority areas of the developer and the priority areas of the consumers. The demand and supply sides, in a sense, are not communicating. And there is certainly a huge market failure when it comes to drugs where those affected do not have buying power.* (governance interviewee, July 2011)

Some participants, such as the multinational pharmaceutical company interviewee and the start-up innovator, argued that patents had limitations primarily in the area of neglected diseases. Several interviewees, such as the civil society and governmental respondents, expressed concerns about general market failure for R&D regardless of type of disease. This diversity reflects their different viewpoints. The private sector representatives work within the patent paradigm as part of their business and view patents as an effective incentive for continuing investment in a project. The civil society perspective is looking at the dilemma from the societal standpoint in which meeting health needs is the main priority, and an innovation incentive should ensure that the new innovation is accessible. The government interviewee is pro-prizes and is concerned with the economic investment of public money in R&D and reducing government costs.

One general example of market failure is that the government is an intermediary, both in the funding of upstream research and the purchase of resulting health products. Both actions of governments mean that the market for health products is not really a "free" market, which by definition means that it is solely at the discretion of private actors without interference from external actors. Increasingly, research that may contribute to the downstream production of a health product is publicly funded; several interviewees felt that the current system of patent protection does not take this into consideration:

> *It ignores the fact that a lot of the research is done outside drug companies, particularly basic research which is done in universities and other public research institutions.* (civil society interviewee, July 2011)

The importance of public health for any country's development and progress means that, as well as funding research, governments are often the primary purchasers of health products. Subsequently, those in need of the health products are often not the primary purchasers. The government perspective argued that this intermediary action of the government means that R&D is not a functioning market. The integral role of the government in R&D is acknowledged by Stiglitz, a Nobel Prize-winning economist whose arguments are quoted by the start-up innovator representative:

> *He [Sitglitz] says unlike open markets in fair competition, healthcare is a national policy-where nations have a very strong reason to take care of health."* (start-up innovator interviewee, June 2011)

A third example of market failure in the general context of R&D of health products is that the demand for a medicine should not be comparable to demand for any other product, due to the absolute need and social externalities associated with medicines.

> *The whole idea is that it is not a market driven logic behind the use of drugs, if you have a disease you need a drug, it is not a need that you can adapt according to your will ... If you are ill you want a cure ... and the need is absolute and not relative ... so you don't have a functioning market.* (government interviewee, June 2011)

The finding from the study is that patent protection may not be a suitable incentive for diseases of the poor, as there is clear market failure in this area. This finding is in line with published research on medicines for diseases of the poor (Trouiller et al. 2001).

## Patents Stimulate Profit-Driven Innovation

Several interviewees acknowledged the fact that patent protection encourages R&D where the most profits can be made. It was identified by the civil society, government, and legal representatives that the priority areas for research may not be in areas where there is the most impact on public health.

> *The patent itself has the incentive of finding medicines that can make the most money, rather than medicines that are in great need or that have therapeutic value. Sometimes these coincide but it means that there are products made to maximize profit that don't have [significant] public value.* (civil society interviewee, July 2011)

> *One of the major problems is that when you look at the research priorities, they are sort of a few key areas. You have anti-ageing research, cancer research and then you have research related to lifestyles so obesity and mental function. Diseases that*

*are ... fundamentally about disease profiles that are chronic, that require people to take pills for a long period of time and that is your dream as a pharmaceutical company, which is an entirely rational dream. But it means that the focus of your research is off in that direction.* (legal interviewee, June 2011)

## Perception of the Need for Alternatives

Nearly all interviewees (except the administrative perspective, who had no opinion on the matter) acknowledged the need for alternatives to a certain degree; however, there was wide variation on the appropriate extent and application of alternatives. The link between patent protection and innovation is well-established, and interviewees identified this deep-rooted link as a potential factor in determining the willingness to change the system. While patents are one form of reward, they are not the only method of rewarding R&D efforts. The government interviewee felt that there is a need for education to break the "mental bond" between patents and innovation in order to open up societal awareness to other possibilities of R&D incentives.

It was expected, due to their dependence on the current system of patent protection, that representatives of the private sector may be completely opposed to the idea of other incentives. However, both representatives acknowledged the need for diverse incentives to stimulate innovation that meets health needs.

*I think that we need different solutions for different diseases and different problems.* (multinational pharmaceutical interviewee, July 2011)

*I'm in the middle of this whole thing, but having said that I already created, in the shareholders' agreement of my company, the provision that for noble causes or real needs, societal needs, we have the provision that we can work on a not-for-profit basis. So yes there is a very big need to actually drive down the cost of healthcare.* (start-up innovator interviewee, June 2011)

Other interviewees corroborated that there is a definite need for other mechanisms in order to stimulate R&D that reflects global health priorities.

*I think that it is definitely necessary, this is not to say that patents are bad actors, but simply it doesn't make sense for them* [pharmaceutical companies] *economically to do this kind of research* [needs-based] *because as much as they would like to be ,they are not interested or not capable and effectively have to be rational. Where they* [pharmaceutical companies] *should be doing research, where we would like them to be doing research, they are not. And so and again, they* [pharmaceutical companies] *are entirely rational, so what we do need then, is to try to find ways to get these companies to do the research themselves.* (legal interviewee, June 2011)

## Should an Alternative be a Replacement or Complement?

With the vast majority of perspectives agreeing that an alternative incentive is needed, the next burning question is whether this alternative should replace the existing system or act as a complement. Some perspectives felt that there is a need for wider change; however, overall it appears from the results that it may be easier and less controversial to introduce alternatives as a complement to the existing system. According to the global health interviewee, a complete overhaul is likely to be met with opposition, would *"raise many questions,"* and would have huge implications for the entire structure of R&D.

An incremental alteration of the system is likely to be more acceptable to key stakeholders, as there were varying opinions on the current system's effectiveness in areas other than diseases of the poor. The interviewee from the multinational pharmaceutical company agreed that patent protection may *"not be the most efficient"* method in the area of neglected diseases; he also felt that there was insufficient evidence to overhaul the existing system for a total replacement, as he argued that it does work well in other areas. The possibility of coexistence of multiple different incentives could solve this problem, as recognized by the governance interviewee:

> *You can either go for a totally needs based research and development model, or entirely patent based, which is profit driven or you can also go for a mix.* (governance interviewee, July 2011)

## Delinking as an Alternative Approach to R&D

This study found that most interviewees, despite their diverse backgrounds, indicated that delinking is a good characteristic of R&D incentives. Several felt that it may be difficult to create in reality but that the concept would promote access to medicines for the poor.

> *Delinking R&D cost from price of medicines, for those living in poverty who cannot access medicines that they need, would be one promising way to address their needs and protect their right to availability of medicines.* (human rights interviewee, June 2011)

Delinking was described by the governance representative as a way to overcome the pharmaceutical industry's justification for high prices; if these two paths were separated, one of the most-used arguments for high prices would no longer apply. Despite strongly supporting delinking, she did point out the need for mechanisms to ensure appropriate management of the products of R&D.

> *Delinking the cost of R&D from the actual price of the final product, and having a price that is simply the marginal costs of production is a good idea ... if you could dissociate the development of production, so everything becomes generic, that would be a fantastic prospect ... But the question is whether this would then stimulate the further development of products and you know, who is going to commercialize the product and get the distribution.* (governance interviewee, July 2011)

The commercialization, production, and distribution were areas where the government interviewee felt that market forces could contribute effectively:

> *Well delinking would be the obvious effect of a prize fund, and that actually puts the possibility of where the market can operate ... in the production of the drugs.* (government interviewee, June 2011)

While it is relatively straightforward to support the conceptual idea of delinking, the challenge is in identifying the specifics required to realize that idea and creating a functioning R&D process in which developing costs and production prices are separate. It is important to note that both interviewees from the private sector thought the concept of delinking was "*plausible*," but that more work needed to be done on specifics. The interviewee from the legal perspective highlighted this dilemma by emphasizing the need to establish how such a paradigm shift could occur; he remained optimistic that it is possible, particularly in the area of neglected diseases.

> *So the basic principle is how do we divorce the incentive to do the research from the price? How do we ensure that the research [is] done and the products produced?... all you have to is say OK ... we will pay you to do the research but you won't get to own the Intellectual Property as a premium on top of that.* (legal interviewee, June 2011)

Overall, delinking was seen as a desirable quality of an incentive for R&D for diseases of the poor.

## A Varying Range of Knowledge of Prizes

Prizes are at an early conceptual stage in the field of R&D. One of the most striking findings was the lack of universal awareness of "prizes" among the interviewees. It is important that all stakeholders are aware of potential alterations to a complex societal system to ensure that the decisions made reflect the wishes of all those affected. While in-depth awareness of patents and innovation was found across all parties, there was a varying range of knowledge on prizes. The perspectives, which were more patient-focused, had greater awareness of prizes (civil society, global health, governance, and government) than those more involved with the supply side of the R&D framework (private sector and administrative).

*Not hugely familiar ... I'm not hugely familiar with other models for R&D ... I am more of a generalist.* (multinational pharmaceutical interviewee, July 2011)

The limited knowledge of prizes among the two private sector representatives is an interesting finding, given their roles within pharmaceutical R&D. It could be argued that as these interviewees work in the current system, they may be most directly affected by adoption of prizes. Furthermore, it was found that other interviewees felt that opposition was most likely to come from the private sector; therefore, informing these actors from the beginning is important to improve acceptability. The finding that interviewees preempt pushback from the pharmaceutical industry is useful; this concern could help mold the specification of prizes to lessen resistance and increase support.

The comparison of the standardized in-depth level of knowledge of the existing system of patent protection and the diversity in levels of awareness of prizes between perspectives implies that prizes are at the early stages of social conceptualization.

## The Potential Place for Prizes in R&D

The concept of prizes is in the early stages of development; now is the ideal time to consider their potential role in R&D. Most interviewees identified the potential worth of prizes in stimulating R&D for neglected diseases, as there is currently a "gap" in R&D that needs to be filled. Some interviewees felt that prizes offer only one choice in a range of alternatives to improve research in this area and would perhaps work best in synergy with other funding mechanisms:

*A mix of prizes and public funds would work really well for neglected diseases, because it would be a substitute in the area where there is no research and development happening at all.* (legal interviewee, June 2011)

In contrast to the common view on the use of prizes for neglected diseases, there was disparity between opinions on the role of prizes for other diseases.

The multinational pharmaceutical company representative proposed that prizes would work best purely in areas in which there are no western markets, such as neglected diseases:

> *I think that these models are almost easier if it is a straight neglected disease without a western market. So if there is a gap in the market. Because I think that if you have a disease that is more mixed then it is more of a burden in a south, but it is also a burden in the north. I think that it really becomes quite difficult then, in terms of making that adjustment in valuation.* (multinational pharmaceutical interviewee, July 2011)

The reason for preferring prizes in the context of medicines for neglected diseases is attributed to the necessary compensation for releasing a patent. Clearly, when there is no market—as in neglected diseases—it is easier for the private sector to engage, compared to diseases for which there is a potential market in developed countries. In opposition, other people's perspectives, such as the civil society and government interviewees, reflected the view that prizes could and should be applied more generally to medicines for diseases affecting both developing and developed countries, such as chronic diseases; prizes could be used to redirect R&D away from profits toward health needs.

## "Prizes" Could Stimulate Needs-Driven Innovation

One main benefit of prizes is their use in "redirecting" research. If prizes were used as an incentive, needs could be identified by global health professionals and a "prize" offered to stimulate R&D in this area. Several interviewees (civil society, government, legal, and start-up innovator) acknowledged the potential benefit of using prizes to steer research toward priority health needs:

> *You can put a focus on where the problems are biggest if you want to deal with neglected diseases, you can put a focus on that, if you want to deal with new antibiotics, because we don't have many, than you can focus on that. So you can help focus the research to the most pressing societal needs, health needs.* (government interviewee, June 2011)

The legal representative emphasized that prizes do not necessarily have to be about the creation of a completely new drug; they may be used to encourage much-needed incremental innovation. Prizes may not be appropriate for completely new medical discoveries that are not planned. One famous example is the accidental discovery of penicillin in 1928 by Sir Alexander Fleming, which is now the most widely-used antibiotic in the world. Sir Fleming is reported to have said, "One sometimes finds what one is not looking for" (Eickhoff 2008). Public grants for basic scientific research are needed to ensure freedom of discovery, with prizes likely to be of benefit further down the development process of new medicines.

## Gaining Support for Prizes

When asked about attitudes toward prizes among their peers, some interviewees felt that there was substantial support, while others identified a significant knowledge gap. Education about prizes was highlighted as the first step needed to gain support. It was noted by several respondents, including the start-up innovator and the government perspective, that despite a lack of awareness, their peers often appear open to prizes when the idea is explained. However, it is hard to predict if they would actively engage in a change to a prize system.

> *I don't hear too many people talking about it* [alternative incentives for R&D for diseases of the poor], *but when I mention it—everybody is sympathetic to it. It's very hard to be unsympathetic about it. If they would actually actively pursue it, or if they would carve out part of their research group to actually do that* [prize based R&D for diseases of the poor] *I don't know.* (start-up innovator interviewee, June 2011)

Breaking the mold of an established system such as the R&D framework always takes time; however, some interviewees, particularly those working in civil society and global health fields, felt that there has already been significant improvement in awareness and discussion of prizes in the past decade. The civil society representative felt that there was significant awareness of prizes in their community; this is reflected in the fact that much of the background literature surrounding 'prizes' is advocacy material from NGOs, and some of the prize proposals have been designed by NGOs (Wilson 2010; KEI 2008). The global health representative also felt that there has been an increase in openness toward prizes over the past decade.

> *I think you know it is gaining quite a lot of support from various key actors and it is being used beyond medicine, so there is a lot of potential in the area.* (civil society interviewee, July 2011)

## What Should the Ideal Prize Look Like?

With perceived increasing support for prizes, there is a need to discuss the preferred qualities of a prize. This study does not discuss the ideal characteristics of a prize for R&D in great detail, but highlights suggestions for prize design, size, funding, and governance.

There are many models of prizes; however, an arbitrary distinction into two main designs can be made. The first type is described by Jamie Love as a "prize fund of a fixed size that would reward drug developers on the basis of the evidence of the impact new products had on health care outcomes" (Medecins Sans Frontieres 2008). This model aims to realign R&D with priority health needs and encourage new drugs that have a significant health benefit over existing treatments. It works on the premise that the R&D process continues as normal, but at the point of approval, a prize is awarded in exchange for the exclusive right to that product.

The second design is based on prespecifying and creating provisions for a medicine that is needed for a population; anyone meeting these specifications within an agreed period is eligible to win a predetermined amount of a prize. An example is the Bangladesh, Barbados, Bolivia, and Suriname (BBBS) Proposal for Chagas Disease, where the prize fund is proposed at $250 million or more. This proposal has several specifications including rewards for openness and sharing between researchers (WHO 2009). This example shows how specifications can be used to modify the R&D process.

When asked on their views toward prize design, most interviewees agreed both have merits, but that the health impact idea may prove challenging for implementation. It would require a complex and robust governance system to effectively measure impact, and there were varying opinions on how easily this could be achieved. The multinational pharmaceutical interviewee argued that it is unfair to reward drug developers on the basis of health impacts when there are many external factors, such as doctor prescribing and environmental impacts, to consider:

> I think that where you move to a more patient-impact orientated decision making it becomes a bit difficult for the industry, because the industry doesn't really supply products direct to patients. Because it is the physician who is responsible for the patient and makes the decision around the product he prescribes. So it is quite hard to put us on the hook really, for overall patient impact. (multinational pharmaceutical interviewee, July 2011)

Interestingly, the interviewee from the governance perspective used the same argument to argue in favor of the health impact-based prizes. While agreeing that it is difficult to measure impact due to the externalities, he opined that it would be a positive step toward evidence-based reimbursement. He highlighted that the additional accountability placed on the industry could be beneficial for patients and governments; companies would start taking an interest in patient compliance and environmental factors, which may lead to improved health outcomes.

*So how do you measure impact is a big challenge, especially when you are trying to measure impact prospectively in the real world setting, where there is a lot of confounders … The impact assessment is attractive because it is effectively talking about evidence-based reimbursement.* (governance interviewee, July 2011)

There is a clear divide in opinions regarding health impact rewards, and prizes that are predetermined may be more appealing, particularly to the private sector. Nonetheless, no interviewee who had a view on this issue dismissed either idea as unworkable.

In addition to two designs of prizes, there are two known ways of rewarding the prize. In order to encourage participation, it was felt by most interviewees that a prize that was rewarded at each stage of the R&D process, i.e., "milestone prizes," would be a more appealing approach than "end stage prizes"; the former design would reduce risk along the R&D process.

*And really with prizes you need to chunk it up into smaller steps. Milestone prizes make sense because when you look at the development of a chemical entity into a patent there are milestones all along that process, so if you can essentially take that process and just offer prizes at each one of those levels then you can really I think do a much better job.* (legal interviewee, June 2011)

One of the biggest dilemmas in the discussion of feasibility of prizes was valuation. What size of prize is fair compensation? Determination of prize size was felt to be very difficult because it is so different from the existing mechanisms of market monopoly. As highlighted by the legal interviewee, the aim of prizes would not be to offer the same monetary reward as a patent because it is a different form of remuneration that is guaranteed and up-front.

The perspective of the multinational pharmaceutical participant gave some interesting insights into the likely thought processes that actors in the industry would consider when deciding whether a prize was worth the investment. Among these are predicted costs and the stage of current science and knowledge as a way forward in the research. Easier economic valuation was one of his strongest arguments for his belief that prizes should be confined to diseases of the poor when there is no potential market to consider in developed countries.

Lastly, sources of financing were discussed. Public funding was identified as the obvious source of financing for prizes. Most of the interviewees believed that this was the most sustainable source of funding and best used in R&D for priority health needs. The interviewees felt that the funding for prizes and other alternative incentives for R&D should be coordinated at an international, rather than national, level. It was also felt that an international governing body would be the most effective way to ensure appropriate prioritization of global health needs and to monitor progress. This finding is in line with recent global developments in R&D, with the CEWG report recommending that WHO member states launch negotiations on a global, binding R&D convention for diseases of the poor (WHO 2012).

## Conclusion

Comments of the interviewees suggest that the current standard incentive mechanism, patent protection, has limitations in the context of diseases that primarily affect the poor; there is a perceived need for alternative (complementary) incentives to stimulate R&D and access to drugs that would work in a context of poverty. These alternatives should be attuned to the specific needs of the poor and to the conditions of the health care systems that serve them. Prizes were not known in great detail by all participants. However, when the core concepts, such as delinking, were explained, there was a general feeling that prizes may have significant potential as an alternative incentive. Prizes were identified as a way to stimulate needs-driven innovation, and the delinking property was seen as a promising way to promote access to medicines. The interviewees identified the need for more discussion on the practical implementation of prizes, as most prize ideas are in the proposal stage. Several interviewees identified the need to start testing prizes and felt that the best area to pilot prizes would be in neglected diseases. This is based on the common view that patents are not working in this area and prizes could fill this gap.

We need to work on finding common ground and act now to further efforts on prizes as an alternative R&D incentive. The past decade has seen many important advances toward a more equitable global society, from the inauguration of the first African-American President of the United States of America to the recent civil uprisings against totalitarian regimes in the Middle East. Now, therefore, is the opportune time to consolidate efforts in the promotion of R&D that meets priority health needs of vulnerable groups.

We need to educate those involved in the R&D process to raise awareness of alternative incentives for innovation, which should be adjusted to specific groups. Therefore, the next step is to initiate a pilot prize in the area of diseases of the poor and to begin global discussions of its results among a wide stakeholder base. Promotion of prizes as part of a wider range of different R&D incentives could help realign R&D with global health needs and provide much needed medicines for those living in poverty. However, as conditions, opportunities, and other context-specific dimensions differ across the globe, one should be careful not to develop a new standardized approach that may work in some contexts, but not in others.

## Notes

1. "Neglected Diseases" do not have a single, preferred definition. One definition is a group of tropical diseases that are especially endemic in low-income populations in developing regions. "Neglected Diseases" are examples of Type 3 Diseases, which are diseases that primarily affect developing countries. They are not synonymous with diseases of the poor, which include Type 2 and Type 3 Diseases. Type 2 Diseases are diseases that occur in both developed and developing countries. Although there are differences, for the purpose of this study, "neglected diseases" and "diseases of the poor" are used interchangeably (WHO 2006).
2. I am very grateful to the nine interviewees who kindly participated in this research and shared their valuable insights on the complex R&D process.

## References

Association of the British Pharmaceutical Industry. 2011. "The Development of New Medicines." Accessed August, 29, 2011. http://www.abpi.org.uk/industry-info/knowledge-hub/randd/Pages/new-medicines.aspx.

Consultative Expert Working Group (CEWG). 2011. "Summary of 2nd CEWG meeting by Chair and Vice-Chair." Presented at Open Session, CEWG on R&D Financing and Coordination, Geneva, July 8.

DiMasi, J. A., R. W. Hansen, and H. G. Grabowski. 2003. "The Price of Innovation: New Estimates of Drug Development Costs." *Journal of Health Economics* 22: 151–185.

Drugs for Neglected Diseases Initiative (DNDi). 2011. "Financing and Incentives for Neglected Disease R&D: Opportunities and Challenges." Comments to the WHO Consultative Expert Working Group (CEWG) on Research & Development: Financing and Coordination. Accessed May 28, 2011. http://www.who.int/phi/news/phi_3_DNDi_submission_CEWG_190611_en.pdf.

Eickhoff, T. 2008. "Penicillin: An Accidental Discovery Changed the Course of Medicine." *Endocrine Today.* Accessed April 12, 2013. http://www.healio.com/endocrinology/news/print/endocrine-today/%7B15afd2a1-2084-4ca6-a4e6-7185f5c4cfb0%7D/penicillin-an-accidental-discovery-changed-the-course-of-medicine.

Griffenhagen, G. B., and J. H. Young. 2009. "Old English Patent Medicines in America." Accessed May 28, 2011. http://www.gutenberg.org/files/30162/30162-h/30162-h.htm.

Hirsch-Kreinsen, H., and D. Jacobson. 2008. *Innovation in Low-Tech Firms and Industries.* Edward Elgar Publishing Limited.

Hotez, P. J., D. H. Molyneux, A. Fenwick, J. Kumaresan, S. E. Sachs, J. D. Sachs, et al. 2007. "Control of Neglected Tropical Diseases." *New England Journal of Medicine* 357(10): 1018–1027.

Hubbard, T., and J. Love. 2004. "A New Trade Framework for Global Healthcare R&D." *PLoS Biology* 2(2): 52.

King, D. M., and M. J. Mazzotta (2000). Glossary." Ecosystem Valuation. Accessed May 22, 2011. http://www.ecosystemvaluation.org/glossary.htm.

Knowledge Ecology International (KEI). 2008. "Selected Innovation Prizes and Reward Programs." KEI Research Note 2008, No. 1. Accessed May 28, 2011. http://www.keionline.org/miscdocs/research_notes/kei_rn_2008_1.pdf.

Knowledge Ecology International (KEI). 2010. *The Baffling WHO EWG Analysis of Innovation Inducement Prizes* (blog). Accessed March 30, 2011. http://keionline. org/node/750.

Love, J. 2011. "De-Linking R&D Costs from Product Prices." Accessed May 27, 2011. http://www.who.int/phi/news/phi_cewg_1stmeet_10_KEI_submission_en.pdf.

Medecins Sans Frontieres. 2008. "The Prize Fund Model: Interview with KEI's James Love." Accessed March 22, 2011. http://www.msfaccess.org/main/medical-innovation/wanted-big-ideas-for-new-drugs/the-prize-fund-model-interview-with-kei-s-james-love/.

Oxfam International. 2008. "Ending the R&D Crisis in Public Health." Oxfam Briefing Paper No. 122. Accessed April 22, 2013. http://www.oxfam.org/sites/ www.oxfam.org/files/bp122-randd-crisis-public-health.pdf.

Patton, M. Q. (1980). *Qualitative Evaluation Methods.* Thousand Oaks, CA: Sage Publications.

Salazar, S. 1998. "Intellectual Property and the Right to Health." Consultant speaking at Panel Discussion on Intellectual Property and Human Rights, Central American Economic Integration Secretariat (SIECA), San José, Costa Rica, Geneva, November 9.

Smith, R. D. 2006. "Responding to Global Infectious Disease Outbreaks, Lessons from SARS on the Role of Risk Perception, Communication, and Management." *Social Science and Medicine* 2(12): 3113–3123.

Stiglitz, J. 2007. "Prizes, Not Patents." *Project Syndicate.* March 6. Accessed March 22, 2011. http://www.projectsyndicate.org/commentary/stiglitz81/English.

Trouiller P, P. Olliaro, E. Torreele, J. Orbinski, R. Laing, and N. Ford. 2002. "Drug Development for Neglected Diseases: A Deficient Market and a Public-Health Policy Failure." *Lancet* 359(9324): 2188–2194.

Trouiller P., E. Torreele, P. Olliaro, N. White, S. Foster, D. Wirth, and B. Pécoul. 2001. "Drugs for Neglected Diseases: A Failure of the Market and a Public Health Failure?" *Tropical Medicine and International Health* 6(11): 945–951.

UN General Assembly. 2008. The Right of Everyone to the Enjoyment of the Highest Attainable Standard of Physical and Mental Health: Report of the Special Rapporteur, Paul Hunt. U.N. General Assembly, 63rd Session, Agenda Item 67(b). U.N. Doc. A/63/263. Available: Accessed 25 May 2011. http://www.essex. ac.uk/human_rights_centre/research/rth/index.aspx.

Wilson, P. 2010. "Giving Developing Countries the Best Shot: An Overview of Vaccine Access and R&D." Report of Oxfam International and Medecins Sans Frontieres Campaign for Access to Essential Medicines, Geneva. Accessed April 20, 2012. http://www.oxfam.org/sites/www.oxfam.org/files/giving-developing-countries-best-shot-vaccines-2010-05.pdf.

World Health Organization (WHO). 1946. "Constitution of the World Health Organization." International Health Conference, New York. Accessed April 10, 2013. http://apps.who.int/gb/bd/PDF/bd47/EN/constitution-en.pdf.

World Health Organization (WHO). 2006. "Public Health, Innovation, and Intellectual Property Rights." Commission on Intellectual Property Rights, Innovation, and Public Health Report, Geneva. Accessed May 22, 2011. http://www.who.int/intellectualproperty/documents/thereport/ENPublicHealthReport.pdf.

World Health Organization (WHO). 2009. "Proposal by Barbados, Bolivia, Suriname, and Bangladesh Proposal." Chagas Disease Prize Fund for the Development of New Treatments, Diagnostics, and Vaccines. Accessed May 29, 2011. http://www.who.int/phi/Bangladesh_Barbados_Bolivia_Suriname_ChagasPrize.pdf.

World Health Organization (WHO). 2010. "Public Health, Innovation, and Intellectual Property: Global Strategy and Plan of Action." Report of the Expert Working Group on Research and Development Financing, Geneva.

World Health Organization (WHO). 2012. "Research and Development to Meet Health Needs in Developing Countries: Strengthening Global Financing and Coordination." Consultative Expert Working Group on Research and Development: Financing and Coordination Report No 1. Accessed April 20, 2013. http://www.who.int/phi/CEWG_Report_5_April_2012.pdf.

# CHAPTER 9

# The Influence of Intellectual Property Protection on Drug Development for Neglected Tropical Diseases

*Aimée Uwland and David Townend*

## Introduction: The Silent Killers

Neglected tropical diseases (NTDs) are a group of 17 tropical diseases designated by the World Health Organization (WHO). These diseases affect a billion of the poorest people worldwide, which is one-sixth of the world's population. They usually have high morbidity rates but do not often result in death, thereby posing an urgent public health concern. In high-income countries, these diseases affect a small number of patients; this limits the market for drug sales to a very small population and makes it impossible to recover research and development (R&D) costs, making pharmaceutical responses to these diseases commercially unattractive. In 2010, the WHO published their first report on NTDs, suggesting they are also "policy neglected" (Crompton 2010). NTDs are strongly associated with poverty, flourishing in poor environments such as remote rural areas, urban slums, and shantytowns. Sufferers are without financial means to seek treatment, making them invisible. Stigmatization that comes with infection produces a vicious circle: disability that further locks sufferers into poverty. NTDs kill people, but not in the same quantity as malaria, tuberculosis, and HIV/AIDS. NTDs are less prominent and, because they do not travel from their low-income locations, they pose no (immediate) threat to high-income countries (Crompton 2010).

Responding to the "neglect" of these tropical diseases, the WHO identified three key factors that contribute to their burden and require an R&D response: failure to use existing tools; inadequate or non-existing tools; and insufficient knowledge of the disease (Trouiller et al. 2002). However, this agenda remains commercially unattractive; the pharmaceutical industry prioritizes commercially profitable products over social or humanitarian incentives. Most pharmaceutical companies are based in high-income countries and target those markets, which does not meet the need for R&D of enhanced or new treatments for NTDs. Drug R&D is a long, highly technological process (often taking 12-15 years) that requires knowledge, time, and resources; consequently, it is capital intensive. It is driven by market incentives and is an integrated process that involves publicly and privately funded research and development (Dalrymple et al. 2006; Maurer 2006; Nwaka et al. 2009; Nwaka and Ridley 2003). Drug R&D for infectious diseases is generally easier than for complex chronic diseases because of the availability of in-vitro/in-vivo microbial elimination tests that directly determine the activity of an antimicrobial entity (Hopkins, Witty, and Nwaka 2007). However, microorganisms that cause NTDs tend to have complex life cycles and require extra-secure, highly-specialized laboratories that make NTD R&D complex and costly.

For some of the NTDs, treatment and prevention is readily available. Early in the twentieth century, R&D was conducted for pharmaceuticals for many of the 17 NTDs. Countries with disease-endemic colonies required treatment and effective prevention to protect their human and animal populations, and some NTDs were nearly eradicated. However, due to decolonization, disease prevention programs stopped and diseases were able to flourish again (Croft 2005). For other NTDs, drug development is necessary; available drugs have problems related to cost, safety, stability, and effectiveness (e.g., because of increasing antimicrobial resistance) (Mrazek and Mossialos 2003; Nwaka et al. 2009). Since the mid-1970s, R&D costs have escalated, while there has been an overall decline in approved new molecular entities (NMEs) (Dalrymple et al. 2006). Absence of intellectual property (IP) protection in NTD-endemic countries and increasing regulations discouraged R&D for NTDs; few new entities entered the development pipeline during the 1980s (Croft 2005; Nwaka et al. 2009; Villa, Compagni, and Reich 2009). The rise in technological molecular and structural biology research, such as genome sequencing, in the 1980s-90s did not benefit R&D for NTDs until the start of the twenty-first century. Furthermore, funding for the translation of newly gained insights into products for the eradication of NTDs remains a challenge (Nwaka and Ridley 2003). Grabowski stated on drug R&D for NTDs: "The basic problem is insufficient expected revenues associated with the low ability to pay for healthcare in poor countries coupled with the high fixed costs of R&D" (Grabowski 2005). R&D costs of the creation of new medicines are high; it is estimated that the cost of developing one marketable drug is five hundred million to one billion dollars, 75 percent of which is spent after entering preclinical testing (Maurer 2006; Nwaka et al. 2009; Nwaka and Ridley 2003). The pharmaceutical industry focuses on patents as an essential source for recovering the overall costs of successful and unsuccessful products and for realizing profits (Hubbard and Love 2004; Villa, Compagni, and Reich 2009). Regulations for drug safety, however necessary, are an important factor in the high costs of drug research and development. For low-income countries, these costs should be put in a cost-to-benefit ratio in which the nonexistence of medication should be taken into account (Trouiller et al. 2001). The pharmaceutical industry argues that drug development for NTDs is "too costly and risky to invest in," as they expect returns to be low from a market that is unable to afford medication (Trouiller et al. 2002). In light of the high costs of pharmaceutical innovation, pharmaceutical companies have pursued "block buster" drugs, commanding high prices; each of these drugs generates a one- to ten-billion dollar profit per year (Scherer 2004). Without such high-profit drugs, large pharmaceutical companies would go bankrupt (Trouiller et al. 2002). Available data shows that only about 10 percent of drug sales go toward R&D of new products for diseases, which contribute to 90 percent of the global disease burden—the so-called 90/10 gap. Furthermore, huge amounts of drug sales are allocated toward marketing expenses (Hale, Woo, and Lipton 2005; Hubbard and Love 2004; Love and Hubbard 2007; Maurer 2006; Nwaka and Ridley 2003).

Available funding for infectious disease research is primarily being spent on "the big three": HIV/AIDS, tuberculosis, and malaria (Nwaka and Hudson 2006). In addition, sustainability of funding is a serious problem (Croft 2005), due to the limited number of funders and an information gap that makes it difficult to motivate new funders (Moran et al. 2009). Investment in tropical infectious disease research in 2007 was just over $2.5 billion (90 percent of which came from public and philanthropic donors, with only 9 percent coming from the private pharmaceutical industry); about 80 percent of this was spent on "the big three" (Moran et al. 2009).

This situation is of great concern, particularly when considering the human right to health care (UN 1948; WHO 1948). The G8s' Millennium Development Goals (MDGs) have produced global interest in the promotion of innovations for NTDs. Failing to address NTDs will adversely impact the ability to reach MDGs 1 (to eradicate extreme poverty and hunger) and 6 (to combat HIV/AIDS, malaria, and other diseases) (UN 2010; Crompton 2010). Additionally, the WHO Commission on Intellectual Property Rights highlighted the importance of innovative product discovery for NTDs (Crompton 2010).

In this chapter, we will discuss the problems related to drug development for NTDs. We will discuss what incentives can be used to stimulate R&D by distinguishing several "push" and "pull" mechanisms. Furthermore, we will determine what solutions are already in play. A literature review conducted on this topic will be accompanied by the visions of several stakeholders. Literature from peer-reviewed journals was found through searching the Maastricht University database, as well as by using Google Scholar. Search terminology included, but was not limited to, (neglected) tropical diseases, drug development, public-private partnerships, intellectual property, and Trade-Related Aspects of Intellectual Property Rights (TRIPS). Articles were included when discussing drug development for neglected diseases, intellectual property in relation to drug development, and/or financing drug development; non-English documents were excluded. Literature data was analyzed using the grounded theory method and used as foundation for qualitative stakeholder interviews. Nine stakeholders were selected for their relevance within the field of NTDs, as well as for their availability for the survey. They were contacted through e-mail, informed of the goal of the study, and asked for cooperation. The interviews were conducted in February 2011 and between the middle of June and early August 2011. The interviews were completed on the basis that the participants would not be personally identifiable in any publications. The participants were: Participant 1 (P1), an executive from a small/medium-sized pharmaceutical company (interview February 4, 2011); P2, P3, and P4, three patent examiners at a Patent Office (interview August 17, 2011); P5, a partner in a law firm specializing in pharmaceutical industry and other biotechnology issues (interview June 29, 2011); P6, a senior lawyer in a large pharmaceutical company (interview July 1, 2011); P7, an NTD official at the WHO (interview July 21, 2011); P8, an advisor to an non-governmental organization (NGO) (interview July 8, 2011); and P9, a lead professor of a department working in the development of pharmaceutical solutions to NTDs (interview July 4, 2011). In semi-structured interviews (one per stakeholder), conducted by telephone or in person, their opinions were gathered. Questions were asked about how international IP rights influence drug development for NTDs; how and why pharmaceutical companies are responding to the issue; and what they feel can and should be done to improve the situation. Notes were made during the interviews and later reduced through grounded theory analysis to compare different stakeholders' opinions and put them in the light of literature data. We will first continue discussing the major problems that can be distinguished as the causes of the neglect of drug development for NTDs.

## Why is the "N" in NTDs?

Three major problems can be identified, which together are the basis of the "neglect" of NTDs and hinder R&D investment (Grabowski 2005; Villa, Compagni, and Reich 2009):

Low monetary incentives and the absence of a market that is able to pay;
Lack of an effective IP protection system—with the danger (for pharmaceutical
companies) of generic drugs—in many low-income countries; and, finally,
Resource-poor settings: weak infrastructure for distributing and selling medicines—leading to problems of accessibility and lack of solid R&D institutions for carrying out their own research in the field of neglected diseases
(Ridley 2004; Scherer 2004).

### Higher Expenses than Revenues

As explained in the introduction, the commercial model can only work when
a market exists that is rich enough to buy patented products to cover research
expenses. This is failing for NTDs in low-income countries where purchasing power is limited (Maurer, Rai, and Sali 2004; Nwaka and Ridley 2003;
Ridley, Grabowski, and Moe 2006; Scherer 2004; Trouiller and Olliaro 1998).
Advances in molecular biology have brought more insights into the microbes
causing NTDs; however, this knowledge has not been translated into new treatments or diagnostic methods (Trouiller et al. 2002). According to Matlashewski
et al. (2011), this situation continues: "There is no shortage of high-quality
research but a lack of translation of research findings into practical therapeutic
interventions" (Matlashewski et al. 2011; Nwaka et al. 2009). Several causes of
this shortage in translation and lack of funding for R&D of neglected diseases
is mentioned in the literature. Hubbard and Love (2004) stated that the existing profit-driven business model for drug development has led to high prices,
unequal access, and widespread dissatisfaction with drug prices; these issues have
been created because the current model uses a single payment to cover the cost
of manufacturing drugs as well as R&D (Hubbard and Love 2004). The lack of
transparency in pharmaceutical R&D investment is also mentioned (Hubbard
and Love 2004; Maurer, Rai, and Sali 2004). In drug R&D, three different
strategies with increasing financial risks for the pharmaceutical industry can be
distinguished (Nwaka and Hudson 2006; Nwaka and Ridley 2003; Stein 2003):

Low-risk, "me-too" approaches where slight changes are made in existing
drugs, which allows for new patents and new marketing;
Medium-risk projects that "piggyback" the activity of existing chemical
agents testing for tropical disease indications; and
High-risk de novo drug discovery projects seeking new classes of inhibitors
against new molecular targets (activity of new molecular entities).

The innovation cycle of high-risk projects typically consists of the discovery and synthesis of new chemical entities, is followed by the translation of findings into product leads, and ends with their further development

into marketable entities. This system leads to the development of new drugs. Large pharmaceutical companies tend to base their choices for drug R&D investment mostly on the value of the expected revenues generated by the developed product. Drugs revenue is estimated by multiplying the price of the drug by the amount of it sold in the market. It depends on the average income of the disease-affected population, the prevalence of the disease, and the infrastructure for distribution of the final drug. Companies are likely to decide not to invest in drugs when the R&D expenses are higher than the expected drug revenues. NTD drugs cannot command either the price or market volume to compete for R&D investment. P1 acknowledged that drug development choices are based upon which drugs the companies think will sell, which drugs fit the type of company and the company's strategy, and whether the company can afford the development. When there is a chance that invested money will not be recovered, private companies will not invest in drug R&D, unless it is for philanthropic reasons or is good for their image.

### Patents to Protect Inventions

Patents have a long history in the encouragement of invention and product innovation. Patents provide the holder with a monopoly over their inventions for 20 years, ensuring freedom of exploitation, particularly over pricing (Drahos and Braithwaite 2001–2002; Maurer, Rai, and Sali 2004; Trouiller et al. 2001). Patents are jurisdictionally specific; they are limited to the extent of the claims described within the patent and the geographic area in which they are filed. Acquiring a patent to cover Europe would cost a company $20,000–$30,000; to cover the main areas of the world would cost about $250,000. However, this is only a small proportion of total R&D costs. Patents do not guarantee income; they provide holders with a "market lead time." In order to justify the monopoly, the law requires disclosure of the specifications of the invention to encourage reproduction after the patent monopoly expires, or to stimulate competitive, original invention to make the patent obsolete. Patents form an essential legal basis to prosecute copying of inventions during the patent's life. Patent examiners (P2–P4) explained: it is a "*one size fits all*" system that covers all sectors, not just the pharmaceutical industry. This is necessary, as placing boundaries between sectors would create gray areas; the system contains a human judgment aspect and, therefore, must be predictable. P5 adds that the system is not conducive for pharmaceuticals and called it a "*bad tool for rewarding innovation in the field*," supporting this statement by giving examples of the amount of deviations (e.g., the pediatric extension plan). It is an "*odd system*," he continued, "*but these are the rules we have got at the moment*" to compensate and protect pharmaceutical companies for the "*enormous risk*" they take in drug R&D. P1 also found the system to be unfavorable. He felt that it is not conducive for pharmaceuticals; however, they know how to work with it. He would prefer the monopoly period to be longer in order to provide a longer period to recoup investments. However, extending patent length is not a realistic solution; patents cover more sectors than just the pharmaceutical industry. P6 mentioned IP to be "*the lifeblood of the company*" and feels the company, for its existence, relies on acquiring patents and marketing products.

Opinions differ between stakeholders as to what extent the current patent and IP system influence drug R&D for NTDs. P6 found the role of IP in NTD R&D to be marginal because of the lack of commercial return: "(The) *IP system has nothing to do with NTDs because it is built for commercial return that patients affected by NTDs do not provide.*" P7 also said that the main problem of drug development for NTDs is not linked to IP problems and continued by saying that removing property rights will not solve the situation. P8 thought of IP protection as a tool and not the main incentive of pharmaceutical companies for drug R&D; the main incentive is the profit they make. However, it is not a good tool for pharmaceuticals, as it works well for products that are not essential and where people are able to pay; however, when problems of access to medicines occur, IP should be changed to solve these problems of accessibility and increase R&D for NTDs.

Patent systems are not supposed to block research because of the existence of research exemptions. When asked, P9 pointed out that he felt constrained by IP as it sometimes prevents him from gaining access to certain molecules or exchanging information. On the contrary, P9 noted that, "*without IP you do not know what others are going to do with your compound and you do not want people to damage your future product.*" As a scientist, P9 would like to spread "*as much information as possible,*" but the IP system constrains P9 from doing so. P9 "*has to deal with it*" and finds ways to work with or around it, acknowledging that IP is needed in order to "*try and prevent others from hijacking your ideas.*" Patents allow them to make partnerships with pharmaceutical companies and maintain control over the drug development process "*without the intention of making money off the poorest of the poor.*" Additionally, holding patents means that revenues will eventually come back to the university when compounds are found to be effective against forms of cancer or in the veterinary industry, making new research possible. However, when others break their patent monopolies, they do not have the significant funds necessary to protect their IP.

The reason we are considering patents in the light of NTD drug R&D is that all of the countries in the world that want to participate in global trade have been forced to sign the TRIPS agreement in order to join the World

Trade Organization (WTO). They must sign in order to gain access to the world trade market, aid, banking facilities, and finance. This has ensured that patents will be internationally recognized. Building up the infrastructure for a patent office is not cheap, and it seems ironic that low-income countries—low on resources—have to invest in a system in which the benefits to them are highly doubtful in order to gain access to this "free trade market." On the contrary, the TRIPS agreement might be beneficial for NTD drug R&D because countries are no longer allowed to produce generics of patented drugs, therefore enabling pharmaceutical companies to earn back their investments. Under TRIPS, P2, P3, and P4 explained that patents still have to be filed in every region to be in effect. Countries are able, apart from certain standard regulations, to set up some rules of their own. In exceptional cases such as during pandemics, patent regulations can be broken (with compulsory licenses) to protect citizens. Products produced then, however, cannot be exported. The use of compulsory licenses is very limited as countries must negotiate with patent holders before they take action. Low-income countries often do not have the technological ability to break patents and produce products for their populations. Uncertainty remains whether strong patent protection will promote R&D of new drugs for NTDs. Opinions among stakeholders are diverse. P6 felt that TRIPS offers a greater degree of certainty, where a robust IP system is helpful in joint ventures and knowledge transfer in terms of looking where to invest, but does not play a role in drug R&D for NTDs. P8 stated that TRIPS *"will not be helpful in drug development,"* as it hinders technology and knowledge transfer.

### Absence of Resources and Coordination

Low gross domestic product (GDP) and drug sales in low-income countries has led to minimal R&D budgets. Healthcare budgets in these countries are marginal, sometimes as little as two dollars per capita per year, directly reflecting their low GDP. Drugs represent a significant proportion of health care expenses (25–70 percent of total health spending) (Grabowski 2005). Expenditure for drug R&D is close to one percent of the countries' GDP, and one-tenth of the drug sales have to contribute to drug R&D (0.1 percent of GDP), as enforced by trade agreements (Hubbard and Love 2004). Low-income countries are therefore unable to meet the high monetary requirements for drug R&D. Building research institutions requires investments in training, equipment, institutions, and networks, all of which are hindered by the absence of money and infrastructure. Due to the absence of research institutions, there is little knowledge exchange, while free exchange of information drives progress in research (Hubbard and Love 2004; Smith et al.

2004). Another problem is uncoordinated research activities that lead to, among other things, the use of different nomenclature for the same microbial strains in different laboratories, hindering cooperation (Nwaka et al. 2009). As resources for drug development for NTDs are poor, the need for coordination of R&D activities and funding for NTDs have become apparent to prevent competition and ensure products are being developed for resource-poor tropical settings (Mrazek and Mossialos 2003). Existing diagnostics or medicines for the treatment of NTDs are not often optimized for these settings as they require hospitalization (intra-venous drugs, specialists, technicians), cold storage, and/or follow-up care, all of which have limited availability in low-income countries (Hopkins, Witty, and Nwaka 2007; Mrazek and Mossialos 2003). Production of these medicines is expensive due to costly pharmaceutical ingredients or high formulation costs (Moran 2005). The inability of countries to manufacture and distribute medicines leads to high drug prices; they are dependent on import, which causes inaccessibility of medicines. Absence of health insurance systems means that 50–90 percent of drug prices are covered by out-of-pocket payments in many low-income countries (Grabowski 2005; Hale, Woo, and Lipton 2005; Love and Hubbard 2007; Mrazek and Mossialos 2003; Nwaka and Ridley 2003; Villa, Compagni, and Reich 2009). Additionally, drug prices are not set but tend to differ from country to country, or even within countries, as prices are set to target the top 5–20 percent of the wealthiest consumers (Love and Hubbard 2007). It is important to realize that weak infrastructure also hinders the implementation of preventive measures, which are important for the full elimination of NTDs (Mrazek and Mossialos 2003).

Now that we have identified many of the problems at the center of the neglect of the research into these diseases, we will continue to examine several mechanisms for stimulating drug R&D and determine if they could form a basis to overcome said problems.

## To "Push," to "Pull," or to "Prize"—How to Incentivize Drug R&D for NTDs?

To stimulate drug R&D for NTDs, "push" and "pull" mechanisms can be distinguished. Push mechanisms are incentives that affect research inputs and reduce R&D costs, thereby reducing the financial risk of development. Some examples of push mechanisms include tax credits, research grants, investment in clinical trial infrastructure, fast-track procedures, and exemption from registration fees. These mechanisms allow the financer some form of control over product development. However, the problem with push mechanisms is the unknown costs of product development and research expenses, thereby making it difficult to estimate how large of a reward to offer (Maurer 2006; Mrazek and Mossialos 2003; Ridley, Grabowski, and Moe 2006; Villa, Compagni, and Reich 2009). Pull mechanisms reward the final developed product (the research output) via financial returns; a guaranteed market is created. Examples include advanced purchase commitments (APCs), preferential pricing, subsidies, enhanced IP rights, tax credits, transferable patent exclusivity rights, and priority review (Hopkins, Witty, and Nwaka 2007; Matlashewski et al. 2011; Mrazek and Mossialos 2003; Nwaka et al. 2009; Ridley, Grabowski, and Moe 2006; Villa, Compagni, and Reich 2009). The WHO, in their Commission on Intellectual Property Rights, Innovation, and Public Health (CIPIH) report, identify APCs, transferable patent exclusivity, and transferable priority review as the pull incentives with the most potential. However, they state that all known incentive systems are flawed in some way and find the choice between offering push and/or pull incentives problematic (Maurer 2006; Towse 2005).

With APCs, sponsors promise to buy a fixed quantity of the developed drug at a predetermined price when development succeeds. The problem is that actual R&D costs are unknown. Funders do not wish to pay too much for the drugs, while pharmaceutical companies want to recoup their costs. There is a minimum threshold of R&D costs, below which no drug development can take place; there is no maximum, however, and funders are challenged to find a price that is reasonable for both parties (Maurer 2006). APCs link R&D to the delivery of the actual product to the patients (Love and Hubbard 2007); they might be a better solution for diseases that have more donor resources, such as HIV/AIDS, malaria, and tuberculosis. Preferential pricing means that drugs are being marketed at higher prices for people that are able to pay (private clinics, high-income countries) and at lower prices for those who cannot. The problem with this incentive is that for NTDs, it would be difficult to identify a patient group that is able to pay, as these diseases mainly affect the poor. Long market exclusivity periods for NTD drugs might not be the solution to the problem either; during this period, prices remain high. With transferable market exclusivity vouchers, companies can extend market exclusivity of "block buster" drugs. However, this solution poses an ethical problem for patients using these block buster drugs: paying for drug development of NTD drugs that they might never need (Mrazek and Mossialos 2003; Villa, Compagni, and Reich 2009). Ridley, Grabowski, and Moe (2006) proposed that the "priority-review voucher," a variation of the extended market exclusivity voucher that would be given to companies that develop drugs for neglected diseases that gain US (or the European equivalent) Food and Drug Administration (FDA) approval, is better than existing treatments and will be manufactured. The voucher would be transferable so that it can be sold to other companies and/or used on drugs that are expected to make a large profit. Priority-review helps drugs come to the market about half of a year earlier, which means that profits can be made for six months longer before the patent expires. Even though this means faster access to block buster drugs in high-income countries, it also means that drugs for NTDs will be financed by consumers of this block buster drug. Additionally, it is debatable as to whether there is any effect on drug safety due to faster review and approval (Ridley, Grabowski, and Moe 2006; Villa, Compagni, and Reich 2009). P9 felt that incentives, like priority-review vouchers, distort behavior as people respond to the changes in policy. P9 felt that certain things cannot be solved on a profit basis. P8 did not think that this was the way to stimulate drug R&D for NTDs: "*These are not the types of strategies* [this NGO] *is pushing for.*" Priority-review vouchers have been in place for four to five years and have not demonstrated efficacy thus far; furthermore, they take resources away from the FDA.

Love and Hubbard (2007) propose a "prize system": instead of a monopoly on the innovation, drug developers are offered large monetary "prizes" that would be tied to the impact of the invention. They argue that recouping the R&D costs via legal monopolies should be divorced from the actual innovation process, as the current patent system is a compromise. Incentivizing drug companies with a prize system instead of market monopoly would lead to the production of more drugs with higher impacts on health, as the highest "prizes" would be awarded for these drugs (Love and Hubbard 2007). It remains to be seen if such a big change can be made and whether large pharmaceutical companies would accept such a different reimbursement system. P2, P3, and P4 felt that to step away from the current IP and drug development system might lead to the "*catastrophic disappearance of current companies.*" It is essential to keep a flow of medicines entering the market; it is impossible to just "*change the rules of the game and make them* [large pharmaceutical companies] *disappear.*" The following section will discuss what is being done to develop medicines for NTDs. Moreover, we will examine whether any of the above-mentioned mechanisms are part of the solution.

## Solutions Already in Play to NTDs?

Participants 2, 3, 4, 5, 6, and 8 thought that the problem of incentivizing the pharmaceutical industry to invent new medicines for neglected diseases can only be solved by direct funding and investing more money. P2, P3, and P4 stated that "*those who have the cash do not have the incentive, and those with the incentives have no cash.*" But is it really that simple?

Apart from the ethical and equity-driven social responsibility principles to invest in drug development for NTDs, there are reasons of self-interest to find solutions for these diseases. Global warming might cause the vectors to cross borders and spread diseases to high-income nations (Ridley 2004). Further, multinational companies investing in NTD R&D gain an improved reputation and strategically establish themselves in the emerging developing countries markets (Moran 2005). P6 acknowledged this by explaining three reasons to invest in NTD R&D. With the IP system under attack in the light of NTDs, showing that research is being done under the current IP system is a way of protecting it. The other two reasons for investing in NTD R&D are that it is the "*right thing to do*," and also enhances their reputation and protects their image.

Large companies are trying to prevent the situation of ten years ago, when the pharmaceutical industry was viewed as a barrier in the battle against HIV/AIDS, from reoccurring. Massive investments have been made (by the Bill and Melinda Gates Foundation among others), but spending these investments efficiently is crucial (Hubbard and Love 2004; Maurer 2006). P7 felt that the landscape for drug development for NTDs has completely changed over the past ten years due to rise of NGOs and networks for drug development. To date, P7 explained, about 80 percent of big pharmaceutical companies have agreements with the WHO to deliver and/or develop drugs for NTDs. No one is willing to develop drugs for NTDs alone, leading to the solution and "*only possibility*" of sharing the costs in "*alliances*" to ensure production and affordability of future drugs. Since 1975, the WHO has partnered with public and private sector organizations in so-called public-private-partnerships (PPPs); these partnerships have proved to be one of the most significant developments in the fight against NTDs thus far (Nwaka and Ridley 2003; Ridley 2004). The development of these PPPs for several stages of the drug R&D process in neglected diseases has accelerated since the late 1990s; by 2005, 75 percent of NTD drug R&D was performed through PPPs (Moran 2005). For all partners, the PPP can be a beneficial solution to problems they are encountering during the R&D process and are unable to solve alone because of the huge development costs of new drugs (Croft 2005). For large pharmaceutical companies (according to their representative, P6), this makes expenses acceptable to shareholders. Academics are able to gain control over downstream product development, as well as the marketing of the eventual product, and pharmaceutical companies gain control over basic upstream research that leads to an easier translation of basic research into clinical development. PPPs can be funded by philanthropic institutions like the Bill and Melinda Gates Foundation, Wellcome Trust, and the Rockefeller Foundation (Nwaka and Hudson 2006). Legal agreements are required to manage the project funding expenditures and benefits since millions of dollars are involved. All parties need to accept these agreements and be comfortable with one another before they can start working together (Nwaka and Ridley 2003). IP rights issues have to be solved through these agreements, as well as the right to continue the project with another partner if one party steps out of the project prematurely (Nwaka and Ridley 2003).

The eventual marketing of the product can be regulated through preferential pricing agreements or other push and pull mechanisms that eventually lower drug prices. The PPP model has proved to be a highly cost-effective method for generating new drugs and seems crucial in the continued involvement of industry in NTD R&D. Apart from the partnerships for drug R&D, pharmaceutical companies have partnered with international organizations like the WHO for donation and distribution of medicines in developing countries (Villa, Compagni, and Reich 2009).

Nwaka (2003) warned that there is a risk of scattered funding and focus of initiatives with duplication of effort, as well as little consensus on which new products are required (Moran et al. 2009; Mrazek and Mossialos 2003; Nwaka and Ridley 2003; Remme et al. 2002; Ridley 2004). From these challenges, some of the current leading initiatives on neglected diseases were born: the WHO Special Program for Research and Training in Tropical Diseases (WHO/TDR, a network of public institutions from high-, middle-, and low-income countries) and the Drugs for Neglected Diseases initiative (DNDi, supported by Médecins Sans Frontières (MSF)) (Hale, Woo, and Lipton 2005; Matlashewski et al. 2011; Maurer 2006). These organizations gather funds via advocacy for neglected diseases and work together with universities and pharmaceutical companies in PPPs to allocate R&D projects. In addition to having good project management, they offer push and pull R&D incentives and find it important to establish win-win collaborations for all parties involved (Hopkins, Witty, Nwaka 2007). To ensure that drugs with any possible effect against NTDs are not shelved, WHO/TDR (cosponsored by United Nations Development Programme (UNDP) and World Bank) and pharmaceutical companies started to work together in the early-to-mid 1990s. WHO/TDR focuses on the early stages of drug R&D by identifying possible active compounds and gathers knowledge on biological, socioeconomic, and behavioral determinants; it also coordinates the development of new/improved tools, intervention methods, and strategies for disease prevention and control in the field (Butler 2003; Gutteridge 2006; Mrazek and Mossialos 2003; Nwaka and Hudson 2006; Nwaka et al. 2009). The coordinated approach has already proven to be important in several ways by setting priorities and linking upstream research to the practical translation into products. WHO/TDR-industry collaborations have generally been successful (Gutteridge 2006; Remme et al. 2002; Ridley 2004).

DNDi is a virtual research organization that aims to create a global public fund for R&D projects for NTDs. They coordinate R&D activities, management, and funds and work through contract relationships with public and private partners. At their beginning in 2003 (the same year Nwaka warned of scattered funding and R&D for NTDs), DNDi aimed to register six to seven new drugs over a decade for $26 million annually. DNDi spreads their funders; they do not wish to get more than 25 percent of their funding from one source. Their funding is more or less equally spread over private (MSF, Bill and Melinda Gates Foundation) and public (Organisation for Economic Co-operation and Development (OECD) countries) partners (Butler 2003; Hubbard and Love 2004; Maurer, Rai, and Sali 2004; Villa, Compagni, and Reich 2009). P8 explained that their NGO's strategy is to push for more development and public funding and to use this funding in the most efficient way: by developing tools that really respond to patients' needs. Private philanthropic foundations do not have the same responsibilities as public institutions, which is why the NGO of P8 would like to increase public funding: "*It is a question of solidarity; the 'West' cannot close their eyes to what is happening in low developed countries.*"

For reasons mentioned earlier, pharmaceutical companies, such as GlaxoSmithKline, Novartis, and AstraZeneca, have set up their own research facilities that direct R&D facilities toward neglected diseases. There are several academic and public centers being financed to research neglected diseases across the world. These universities work together with the pharmaceutical industry on projects allocated by DNDi or the WHO to develop active compounds into new products (Nwaka et al. 2009; Scherer 2004). The animal health and agrochemical industries have contributed compounds for evaluation for NTDs; the pharmaceutical industry is now involved in nearly half of new NTD drug development on a noncommercial basis, with budgets large enough to deliver a new drug every few years (Dalrymple et al. 2006; Hopkins, Witty, and Nwaka 2007; Maurer 2006; Moran 2005; Nwaka and Hudson 2006).

Finally, several different networks for specific NTD R&D have been established from these initiatives. These networks have a cost-effective and open, public-health centered approach to the discovery of new drugs for NTDs (Nwaka and Hudson 2006; Nwaka et al. 2009). Most of these networks and initiatives are based in high-income countries, but efforts are being made to involve institutions in developing countries. Companies in these nations often fear the risk of poor financial return, but the participation of native researchers is valuable for the exploration of natural products and technology transfer (Nwaka and Hudson 2006). An example is the establishment of the African Network of Drugs and Diagnostics Innovation (ANDI), with the objective of discovering, developing, and delivering new products for African-endemic diseases through promoting and sustaining African R&D. Researchers in disease-endemic areas have the advantage of being able to research natural products and traditional medicines. Additionally, clinical trials can be performed more cheaply in developing countries, lowering R&D costs (Butler 2003; Nwaka et al. 2009; Nwaka and Ridley 2003).

## Conclusion

This chapter discussed the problem of drug R&D for NTDs and how (and whether) the IP system contributes to the "neglect." It became clear that principal incentives to stimulate private sector R&D are monetary, and ensuring social justice is not "the market's" primary focus. As patents do not guarantee income, IP rights (according to P6 the *"lifeblood"* of pharmaceutical companies) play a marginal role when a viable market is absent. NTDs do not pose a threat to high-income nations; those affected by these diseases often live on less than one dollar per day. We saw that drug development is a costly process that is too risky for companies to undertake alone, especially when expected revenues are low. Additionally, the absence of recourses and research coordination has been discussed in the light of NTDs. The absence of monetary incentives in a market unable to pay in a commercial model that has high R&D costs, put against a backdrop of resource-poor settings with an absence of research institutions, uncoordinated research, and inability to manufacture and distribute drugs in disease endemic countries, form the fundament of neglect of these 17 diseases. To stimulate drug R&D, several mechanisms have been discussed in the light of NTDs, but the real road to a solution lays in a comprehensive approach. Funding is centrally gathered and research activities are coordinated. Academic and private partners form alliances and "team up" in PPPs, which gain funding and assignments from central instances. Through PPPs, responsibilities are shared, "push" and "pull" R&D incentives are offered, and win-win collaborations can be established. The formation of networks leads to sharing knowledge, resources, and involvement of stakeholders and makes it possible to work on NTD R&D most efficiently. Reasons for commercial parties to join NTD drug R&D include protecting the IP system (by showing research is being done), philanthropy (*"it is the right thing to do"*), image, and exploring new markets. As most disease-endemic countries are developing, investing now might pay off later. Further, global warming might prove to be a future incentive. When TseTse flies extend their living area to Spain or *Aedes aegypti* spreads over the United

States and Europe, governments of these countries will then readily push for solutions. Until that happens, the gathering of funds via advocacy and the central allocation of R&D projects through PPPs and research networks will continue. Whether this will prove to be sufficient is uncertain; follow-up research on the effectiveness of this approach in combating NTDs is needed.

## References

Butler, D. 2003. "Tropical Diseases: Raiding the Medicine Cabinet." *Nature* 424(6944): 10–11.

Croft, S. L. 2005. "Public-Private Partnership: From There to Here." *Trans R Soc Trop Med Hyg* 99 Suppl 1: S9–14.

Crompton, D. W. T., D. Daumerie, P. Peters, and L. Savioli. (2010). "Working to Overcome the Global Impact of Neglected Tropical Diseases. First WHO Report on Neglected Tropical Diseases." *Geneva: World Health Organization: Department of Control of Neglected Tropical Diseases.*

Dalrymple, M., D. Taylor, C. Kettleborough, J. Bryans, and R. Solari. 2006. "Academia-Industry Partnerships in Drug Discovery." *Expert Opin. Drug Discov.* 1(1): 1–6.

Drahos, P., and J. Braithwaite. 2001–2002. "Intellectual Property, Corporate Strategy, Globalization: TRIPS in Context." *Wisconsin International Law Journal* 20: 451–480.

Grabowski, H. 2005. "Increasing R&D Incentives for Neglected Diseases—Lessons from the Orphan Drug Act." In K.E. Maskus and J.H. Reichman (eds), *International Public Goods, and Transfer of Technology Under a Globalized Intellectual Property Regime, Cambridge University Press, Cambridge,* 457–480.

Gutteridge, W. E. 2006. "TDR Collaboration with the Pharmaceutical Industry." *Trans R Soc Trop Med Hyg* 100 Suppl 1: S21–25.

Hale, V., K. Woo, and H. Lipton. 2005. "Oxymoron No More: The Potential of Nonprofit Drug Companies to Deliver on the Promise of Medicines for the Developing World." *Health Aff (Millwood)* 24(4): 1057–1063.

Hopkins, A. L., M. J. Witty, and S. Nwaka. 2007. "Mission Possible." *Nature* 449(7159), 166–169.

Hubbard, T., and J. Love. 2004. "A New Trade Framework for Global Healthcare R&D." *PLoS Biol* 2(2): E52.

Love, J., and T. Hubbard. 2007. "The Big Idea: Prizes to Stimulate R&D for New Medicines." *Chi.-Kent L. Rev.* 82: 1519.

Matlashewski, G., B. Arana, A. Kroeger, S. Battacharya, S. Sundar, P. Das, P. Kumar Sinha, et al. 2011. "Visceral Leishmaniasis: Elimination with Existing Interventions." *Lancet Infect Dis* 11(4): 322–325.

Maurer, S. M. 2006. "Choosing the Right Incentive Strategy for Research and Development in Neglected Diseases." *Bull World Health Organ* 84(5): 376–381.

Maurer, S. M., A. Rai, and A. Sali. 2004. "Finding Cures for Tropical Diseases: Is Open Source an Answer?" *PLoS Med* 1(3): e56.

Moran, M. 2005. "A Breakthrough in R&D for Neglected Diseases: New Ways to Get the Drugs We Need." *PLoS Med* 2(9): e302.

Moran, M., J. Guzman, A. L. Ropars, A. McDonald, N. Jameson, B. Omune, S. Ryan, and L. Wu. 2009. "Neglected Disease Research and Development: How Much Are We Really Spending?" *PLoS Med* 6(2): e30.

Mrazek, M. F., and E. Mossialos. 2003. "Stimulating Pharmaceutical Research and Development for Neglected Diseases." *Health Policy* 64(1): 75–88.

Nwaka, S., and A. Hudson. 2006. "Innovative Lead Discovery Strategies for Tropical Diseases." *Nat Rev Drug Discov* 5(11): 941–955.

Nwaka, S., B. Ramirez, R. Brun, L. Maes, F. Douglas, and R. Ridley. 2009. "Advancing Drug Innovation for Neglected Diseases—Criteria for Lead Progression." *PLoS Negl Trop Dis* 3(8): e440.

Nwaka, S., and R. G. Ridley. 2003. "Virtual Drug Discovery and Development for Neglected Diseases through Public-Private Partnerships." *Nat Rev Drug Discov* 2(11): 919–928.

Remme, J. H., E. Blas, L. Chitsulo, P. M. Desjeux, H. D. Engers, T. P. Kanyok, J. F. Kengeya Kayondo, et al. 2002. "Strategic Emphases for Tropical Diseases Research: A TDR Perspective." *Trends Microbiol* 10(10): 435–440.

Ridley, D. B., H. G. Grabowski, and J. L. Moe. 2006. "Developing Drugs for Developing Countries." *Health Aff (Millwood)* 25(2): 313–324.

Ridley, R. G. 2004. "Research on Infectious Diseases Requires Better Coordination." *Nat Med* 10(12 Suppl): S137–140.

Scherer, F. M. 2004. *A Note on Global Welfare in Pharmaceutical Patenting.* Malden, MA: Blackwell Publishing Ltd.

Smith, R. D., H. Thorsteinsdottir, A. S. Daar, E. R. Gold, and P. A. Singers. 2004. "Genomics Knowledge and Equity: A Global Public Goods Perspective of the Patent System." *Bull World Health Organ* 82(5): 385–389.

Stein, R. L. 2003. "High-Throughput Screening in Academia: The Harvard Experience." *J Biomol Screen* 8(6): 615–619.

Trouiller, P., P. Olliaro, E. Torreele, J. Orbinski, R. Laing, and N. Ford. 2002. "Drug Development for Neglected Diseases: A Deficient Market and a Public-Health Policy Failure." *Lancet* 359(9324): 2188–2194.

Trouiller, P., and P. L. Olliaro. 1998. "Drug Development Output from 1975 to 1996: What Proportion for Tropical Diseases?" *Int J Infect Dis* 3(2): 61–63.

Trouiller, P., E. Torreele, P. Olliaro, N. White, S. Foster, D. Wirth, and B. Pécoul. 2001. "Drugs for Neglected Diseases: A Failure of the Market and a Public Health Failure?" *Trop Med Int Health* 6(11): 945–951.

United Nations (UN). 2010. "The UN Summit on the Millennium Development Goals." Accessed April 10, 2011. http://www.un.org/millenniumgoals/.

Villa, S., A. Compagni, and M. R. Reich. 2009. "Orphan Drug Legislation: Lessons for Neglected Tropical Diseases." *Int J Health Plann Manage* 24(1): 27–42.

Towse, A. 2005. "A Review of IP and Non-IP Incentives for R&D for Diseases of Poverty. What Type of Innovation Is Required and How Can We Incentivise the Private Sector to Deliver It?" *Geneva: World Health Organization: Commission on Intellectual Property Rights, Innovation and Public Health.*

# Contributor Bios

**Lloyd Akrong** holds a Master of Science in Global Health from Maastricht University. He is currently a Research Fellow in the Department of Health, Ethics, and Society. His interest is in the sociology of biomedicine with a focus on the globalization of clinical trials.

**Inge A.S van Alphen** holds a Master of Science in Global Health from Maastricht University with a minor in Human Rights from Thammasat University, Thailand. For over a decade, she has lived in India, where she now works as a public health consultant on local health projects and teaches biology in her former high school.

**Elena Ambrosino** has a background in Immunology and Molecular Biology and International Health. She is an assistant professor at the Institute for Public Health Genomics (IPHG) at Maastricht University.

**Daniel K. Arhinful** is a Research Fellow in the Department of Epidemiology at the Noguchi Memorial Institute for Medical Research at the University of Ghana. His research interests are in health insurance, access to medicines, and migrant health.

**Angela Brand** is Founding Director and Full Professor of the Institute for Public Health Genomics (IPHG) at Maastricht University, the Netherlands. She also holds the Dr. T.M.Pai Endowed Chair on Public Health Genomics in Manipal University, India.

**Marianne Eelens** holds a Bachelor of Science in Interdisciplinary Studies in Social Sciences: Health Studies from Michigan State University. In 2011, she completed a Masters in Global Health from Maastricht University. She is working at the Netherlands Interdisciplinary Demographic Institute (NIDI) as a Research Associate.

**Nora Engel** is an assistant professor of Global Health at Maastricht University. Her work focuses on innovation dynamics in global health challenges (such as tuberculosis) and on the sociology of diagnostics and innovations at the point-of-care in India and South Africa.

**Klasien Horstman** is Professor of Philosophy of Public Health at Maastricht University and head of the Health, Ethics, and Society Research Program. Her research focuses on philosophical-sociological analysis of public health issues at the interface of science, technology, and society.

**Ine Van Hoyweghen** is a research professor at the Centre for Sociological Research (CeSO) at KU Leuven and a member of the Young Academy of the Royal Flemish Academy of Belgium for Science and the Arts (KVAB). Her work focuses on the societal, regulatory, and ethical dimensions of biomedicine.

**Anja Krumeich** is an associate professor in Global Health and director of the Maastricht Global Health program. In her research, Anja applies insights from participatory research, critical ethnography, and Science and Technology Studies on the interaction between local and global dimensions of health and health care practice and issues of governance in global health in Africa, Latin America, and Europe.

**MeiLee Ling** holds a Bachelor of Science in Microbiology from Oregon State University, USA, and a Master of Science in Global Health from Maastricht University. She is a PhD candidate in the Bioscience Department, Aarhus University.

**Agnes Meershoek** studied Health Care Sciences and obtained her doctoral degree on a study into the daily practices of physicians' illness certification for the Dutch Sickness Benefit Act. She is an assistant professor in Health Sciences at Maastricht University.

**Lois Murray** holds a medical degree (Hons) from Queen's University, Belfast, and a Master of Science in Global Health (cum Laude) from Maastricht University. She is currently working as an academic foundation doctor at King's College Hospital, London.

**Phuong Nguyen Thi Mai** completed a Master of Science in Global Health from Maastricht University in 2012. Since then, she has worked as a program officer for the Vietnam National TB program "Programmatic Management of Drug-resistant TB".

**David Townend** is an academic lawyer. He is Professor of Law and Legal Philosophy in Health, Medicine and Life Sciences at Maastricht University. His research interests include the concepts of property, privacy, the public interest, and politeness in professionalism.

**Mario Vaz** is Professor of Physiology at St. John's Medical College, Bangalore, India, where his work has focused on the functional consequences of altered nutritional states across the nutritional spectrum. He is also Head of the newly founded Health and Humanities Division of St. John's Research institute and curates the Major General SL Bhatia History of Medicine museum.

**Maria M.C. Verhagen** completed a Master of Science in both Medicine and Global Health at Maastricht University in 2012. She is currently working as a community doctor in the field of Youth and Public Health.

**Aimée Uwland** holds a Bachelor of Science in Medicine and a Master of Science in Global Health. Her studies introduced her to work environments in different areas of health care and research; she accepted jobs as a contractor laboratory scientist at Unilever and Cargill respectively.

**Olga Zvonareva** holds a Master of Science in Global Health from Maastricht University and works as a research fellow at the Department of Health, Ethics, and Society at Maastricht University. She is an honorary associate researcher in the Steve Biko Centre for Bioethics at the University of Witwatersrand, South Africa.

# Index